COSMOLOGY AND THEOLOGY

TERRY CAIN

AuthorHouse™
1663 Liberty Drive
Bloomington, IN 47403
www.authorhouse.com
Phone: 1-800-839-8640

© 2014 Terry Cain. All rights reserved.

No part of this book may be reproduced, stored in a retrieval system, or transmitted by any means without the written permission of the author.

Published by AuthorHouse   12/11/2014

ISBN: 978-1-4969-5412-1 (sc)
ISBN: 978-1-4969-5641-5 (e)

Library of Congress Control Number: 2014921537

Any people depicted in stock imagery provided by Thinkstock are models, and such images are being used for illustrative purposes only.
Certain stock imagery © Thinkstock.

This book is printed on acid-free paper.

Because of the dynamic nature of the Internet, any web addresses or links contained in this book may have changed since publication and may no longer be valid. The views expressed in this work are solely those of the author and do not necessarily reflect the views of the publisher, and the publisher hereby disclaims any responsibility for them.

CONTENTS

Chapter 1 In the beginning 1
Chapter 2 Universe 5
Chapter 3 God 43
Chapter 4 Immortality 62

ABOUT THE AUTHOR

Dr. Cain has served most of his ministry in the Nebraska Annual Conference of the United Methodist Church. His education includes the following: • Bachelor of Arts from Nebraska Wesleyan University • Master of Divinity from St. Paul School of Theology, Kansas City, Missouri • Doctor of Ministry from San Francisco Theological Seminary • Extensive undergraduate and graduate work in philosophy, sociology, and geology at the University of Nebraska, Lincoln • Graduate work at Duke University.

ACKNOWLEDGEMENTS

If you detect a Nicaraguan accent in the text, it is the influence of Lucia Espinosa who did major typing for me and deserves much thanks for putting up with my erratic dictation. She remains a good friend.

My other good friend is Marvin Cole, a North Carolina native whom I consider the best Mark Twain impersonator and authority in the whole wide world. The material from Samuel Clemens's that he found for me is very much appreciated.

I suppose I must thank my family (wife, Sue; son, Terry II; and daughter, Sherry Clements) as they did contribute many ideas and much word selection. Besides trying to make sense of my confusion with their amazing conversations, Sue also typed much of the manuscript – Sherry was my agent and got the book published – while Terry, my son offered incredible help as one of the most knowledgable persons in the field of astronomy. The three of them will try to take credit for the best material in the book – if there is any. Don't talk to them!!

I'm grateful to NASA for the photo used on the front cover. NASA's photos are in the public domain.

1

IN THE BEGINNING

"The researches of many commentators have already thrown much darkness on this subject, and it is probable that, if they continue, we shall soon know nothing at all about it."
(Mark Twain)

The prevailing theory (and a good one!) is that you were created inside a very large star! Most of the elements that make up your body – iron, nitrogen, sulfur, hydrogen, carbon, and everything else found in your body – were formed inside a star that eventually exploded and sent debris (including the material used to create your body) spewing out into space only to be gathered together again into another star (our sun) and our solar system including Earth, where we materialized! The process took billions of years, but here you are!! Exciting? Awesome?

A person who enjoys an inquiring mind conceivably has a richer interaction with the universe or at least with the world he or she experiences on a daily basis. Inquiry exercised assiduously can open fascinating possibilities just as one enjoys the anticipation and revelation associated with the opening of a birthday gift. There is no doubt that early life experiences develop that sense of wonder and curiosity which causes us to ask questions and seek knowledge. Personal encounters, street-smart experiences, and perhaps most of all, reading, broadened our world. Today, however, technology beyond our wildest dreams has opened up worlds of possibilities to stretch our understandings and imaginations. A case could be made that the more we know concerning a thing or event the more we can enjoy our experience of that item. Consequently our lives should be enriched and made more enjoyable through our search for answers to life's more important questions.

As you read this book, be prepared to encounter many superlatives, such as amazing, awesome, incredible, marvelous, and other similar descriptions. It is the nature of the subject matter to generate such adjectives. We wouldn't expect to discuss the universe, God and immortality without using strong descriptions.

If you haven't lost interest in the direction of this discussion so far, it should rekindle your interest to know that regardless of whether this book will have anything worthy to say; the major theme, at the least, should capture the imagination of most people. Even if this book does not satisfy your expectations, it shouldn't deter you from pursuing other sources in a search for answers or relevant discussions concerning that major theme. Hopefully, you are one of the inquiring minds that spends some time asking yourself the questions: What is the nature of the universe?; Is there a God?; What is the relationship of God to the physical universe and its inhabitants?; What happens at death? Is there an on-going existence of the mind/soul/spirit or is it simply destroyed – never to exist in eternity?

The theme of this book is to see the above questions, and related ones, as belonging to one great subject of inquiry. The author finds it inconceivable that anyone would not find themselves pondering these ideas on a regular basis. Can anyone not wonder about the mystery of death and what happens to a person after it? Do we not exercise ourselves over the matter because we may be a Christian and feel the New Testament has resolved all questions? Or do we find the subject too frightening to contemplate?

Large numbers of people buy books that appeal to our greed, or rehash the old idea of the power of positive thinking, or simply puts forward ideas we all agree with and welcome as a validation of our prejudices (all of which are happening in inexplicable sales rewarding authors with nothing worthwhile or profound to say). Shouldn't we be far more interested in thinking about our universe or what happens to us after our inevitable death? This is not meant to promote *this* book. It is simply wondering why we do not have a greater interest in the more critical questions concerning life.

The three topics I hope to tie together are the universe, God and what happens to us when we die. I assume these three subjects have to be inseparably related.

Let us set some ground rules to provide the foundation for our discussion. The following axioms will be the basis or framework for the ideas and arguments discussed in this book. They will be repeated in other parts of the book as they become relevant.

One: ***The universe, creation, God's world is logical and makes sense***. Some things about our world seem strange and unreasonable (for example: why a good god would allow pain and suffering) until we unravel them and find there are rational explanations. When we understand what is going on we find that it makes sense. The world is reasonable and sane!

Two: ***Our values are practical and dependable***. Philosophers tell us that we cannot judge physical or moral laws by our own values. For example: we might say that life is cruel and not fair. Philosophers insist we are using our value systems to assess a universe that does not run on our value systems. The universe does not have to concur or resonate with our principles. Our values are based on our limited perspective and lack of knowledge. I want to assert this second axiom as a truth for the basis of our discussion. If we cannot use our value systems, then we have nothing. Even if we do not agree among ourselves on what those values are; however, each of us must work out our own set of principles, regardless of what others believe. If we cannot impose our values on a judgment of the universe or life, then we have nothing with which to measure. Thus we have nothing to talk about. Conversation is over. To make any progress in understanding the world we live in, we have to accept axiom # two as one of our ground rules.

Three: ***Simple is usually right***. The fourteenth century philosophy of Ockham's razor states the following: Given alternate theories or possibilities, the simplest one is usually right.

Four: ***The physical (and moral) laws that pertain here in our part of the universe are universal and apply everywhere.***

I found an axiom in a book on the chemistry periodic table that seems to summarize the four axioms we propose: *"Natura nihil agit frustra"* which translates as "there are no grotesques in nature" which that author considered to be "the only undisputed axiom in philosophy; which is in itself, a questionable assessment. The point being that nature tends to be less strange, odd or bizarre than not. More often than not nature will take the more simple and commonplace route rather than the more complex and weird one.

In order to continue this discussion we need to accept these four axioms as the basic foundation for the ideas of this book, even if they are not personally your own beliefs. Without this foundation for discussion we have nothing to talk about. We are not arguing the truth of these axioms. I begin by assuming them and develop a story that they support.

* * *

Warning: read this book and you will want to read it over and over again. You may become a person with a one book library (three books if you include the dictionary and telephone directory).

2

UNIVERSE

The cover of this book is one of the most incredible photographs ever taken! (The title of the book and author's name did not appear in the original NASA photo.) The picture was taken with the Hubble Telescope over a ten day period of exposures. Such long exposures are necessary to register the light from such faint objects. Each fuzzy spot of light or elliptical object is a galaxy, each one containing more than a hundred billion stars! We're told that the nearest ones are millions of light years distant and many of the faintest ones are billions of light years away! How far is that? Remember if you turn on a flashlight tonight and aim it parallel to the ground, and if the light could bend with the curvature of the Earth and travel around the Earth and back to you again at the speed of light, it would make the trip around the Earth seven and a half times in one second! I am using exclamation marks because these facts are startling and unbelievable. Even at that incredible speed (seven and a half times around the Earth in one second), it takes light from some of these galaxies a billion or more years to reach us!! Yes, two exclamation marks! If this information by itself doesn't intrigue your imagination and excite you, then that in itself is an astounding fact!

Let me share a reason for beginning this book with considerable material describing the nature of our universe. The purpose is to overwhelm you with the magnificence of our universe that you might be impressed with the power and organized complexity that would be questionable to attribute it to random chance.

Take another look at the cover. The electromagnetic waves that resulted in this exposure started out millions and billions of years ago and only recently reached Earth. The nearest galaxy to us is Andromeda at over a million light years away. (It is not in the photo on the cover.)

Astronomy is one of the most fascinating of subjects, if not *the* most fascinating of all. Part of its charm, and also its frustration, comes from the awesome vastness of the universe. Contemplating these aspects of the cosmos is a dizzying exercise. The job of astronomers, physicists and mathematicians is to investigate and then describe the universe. The frustration just mentioned lies in the fact that most of the subject matter is so far away and many present too many obstacles for accurate examination and measurement.

Scientists pose theories to explain the mysteries of black holes, quasars, dark matter, and other objects as well as simply the overall nature of the entire universe. Anyone familiar with the subject knows that the theories, descriptions or ideas put forward seeking to explain the more exotic elements of the universe inevitably fall short of a satisfactory explanation and leave some embarrassing contradictions unaccounted for. For example, one noted theoretical physicist admitted that his colleagues are "well known for the prodigality with which they invent and discard theories." He admitted there are myths and mysteries in cosmology that create confusion and inaccuracies.

So can a non-scientist offer some ideas concerning the anatomy of the universe? If there are some stark inconsistencies or unanswered problems in such amateur ideas, how would that be different from the current and less than adequate theories of the professionals regarding the Big Bang, the expanding universe, String theory, the curvature of space, multiple universes and so forth?

The thoughts in this chapter tend towards the old fashion Newtonian laws of physics as a foundation for some new reasons for revisiting old concepts. An example would be an infinite universe that is not expanding despite what the red shift Doppler Effect suggests. I also argue that space does not curve. We are not the center of the universe. And perhaps much more.

Not being a scientist, the following suggestions may seem amateurish. I warn you that I am a theologian. However, this means that I will try to get away with claiming that instead of getting my ideas on the universe from scientific observation and experimentation, I get my information directly from the original source. (And "God" wouldn't lead me astray would "he"?) That should eliminate any rebuttal. Nevertheless, I will simply label the proposals in this chapter "thoughts" and not theories.

Disclaimer: These thoughts belong to a non-scientist amateur and reflect the influence of an old fashion astronomy. Some of the discussion is much out of date; some is current and acceptable, and some is "right on" and accurate despite being dismissed by and drawing the ire of scientists.

What is the purpose of this chapter? It would seem that any discussion concerning the nature of God should include an examination of God's creation, the universe – what we think we know of it so far. Having said that, it would seem to follow logically that every theologian or person with a desire to study God should become a mini-scientist. A theologian should have an insatiable desire to know more about the world and the universe either by reading and study or personal pursuit. A religious person should be filled with cosmic wonder.

Some noted scientists started their careers as theology students. Indeed, Darwin began his studies in theology. One of the reasons he finally became more of an agnostic was due to his observation of the cruelty and violence of nature. I will try to take some of the sting out of this problem later in this book.

Isaac Newton wanted to know the mind of God and so he pursued serious religious studies. The scientist and Dominican priest, Giordano Bruno, was burned at the stake in 1600 AD for claiming the sun was just one of many other suns (stars) in the universe. Copernicus was a canon of the Roman Catholic Church. The list goes on.

Now we come to the main theme of this chapter. Next to the difficulties of measuring a subject so vast, the Christian church was one of the serious obstacles to cosmological advances. A major controversy focused on the religious tenant that we, the Earth, were at the center of the universe and everything else. If it (stars, planets, moons) moved, it moved around the Earth. Early Christian theology saw it as an imperative to have humanity be the centerpiece of God's creation - the navel. Consequently, the various theories of an alternative center of our universe, such as the sun, would be heresy. As an example of the pervasiveness of theology, the second century AD astronomer, Ptolemy, gave us a contorted arrangement of circles to account for the observable movement of the planets around the Earth in order to conform to the church's teaching. In the 16th Century, Tycho Brahe tried to correct the deficiencies of Ptolemy's explanation by having the planets move around the sun, but keeping the Earth in the center by having both the sun and planets move around us.

The two early cosmologists who dared take the Earth out of its rightful religious position as center of the universe were Aristarchus and Copernicus. This was disturbing to the church. The church said the Earth had to be the center of the universe. Science finally came to the position that, no, the Earth is not the center of the universe. The church did not give up. Galileo (a seventeenth century astronomer who was also guided by our third axiom as he asserted that the laws of nature are simple, not complicated) was condemned by the Roman Catholic Church for teaching that the Earth revolved around the sun and was not the center of the universe.

The point of this discussion here is the ironic turn of events that may prevail today. From all the discussions of cosmologists that I read, the current theories actually place the Earth back in its original spot as the center of our universe. I, as a theologian (confuting the church's traditional position), believe we are not the center! Let me explain.

Astronomers look out in every direction from the vantage point of Earth and perceive galaxies as far as the telescopes can see. Using their difficult attempts to measure the distance to those galaxies furtherest out, they determined they are roughly 14 billion light years away. They also assume they may be observing the edge of the universe and that there are apparently no more (or very few) galaxies beyond this point. Since this observation is true in every direction, they apparently believe we are once more at the center of the universe! Since they believe they are viewing the outer limits, the age of the universe is determined to be 14 billion years old. More on the problems this assumption creates later.

Figure 2.1 represents a slice through the middle of the universe, if we visualize it as many astronomers do, as an expanding structure analogous to a balloon being blown up. The circle with the "E" is Earth (or more precisely, the Milky Way galaxy containing Earth and our solar system) and the other objects are galaxies. The large circle represents the edge of the universe as proposed by astronomers today.

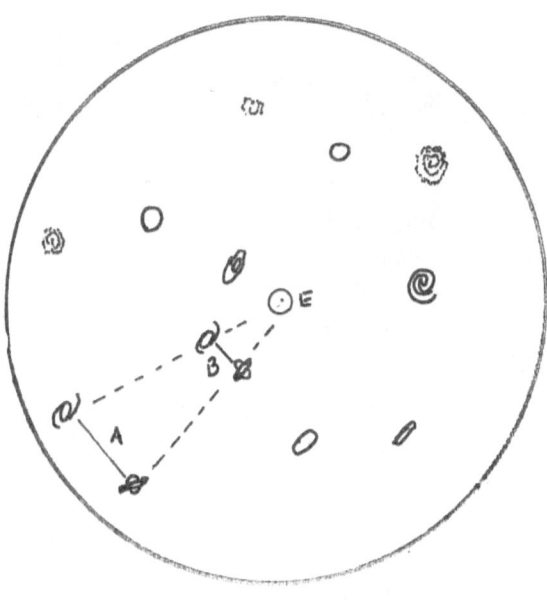

Figure 2.1

Most scientific discussion is carried on as if we are more or less at the center, and three arrangements concerning the location of galaxies emerge. One is illustrated by figure 2.1 with the run-away galaxies expanding or moving away from us distributed on the outer edges, leaves the central region devoid of galaxies due to their vacating the area in their move away into outer space (which isn't considered space at all in most models). The universe according to some astronomers is not expanding into empty space, but instead into nothingness which isn't space as you and I think of it. Simple observation indicates this isn't the case. Galaxies fill all regions somewhat uniformly, exceptions are negligible and not relevant to this discussion. If space and the universe were expanding, it would seem to be the case that the galaxies "farther out" from the center would on average be further apart than those not as far out towards the edge. Compare lines A and B in figure 2.1 to see this. Yet actual observations of the spacing of galaxies do not bear this out as I understand it to be the case.

A second suggestion would have all matter (stars and galaxies) being continuously created near the center in order to replace the ones racing away as they vacate the central region. If not there would be a central void that was continually expanding. And where is that central region?

The most popular theory, I think, postulates an expanding universe where all the galaxies have, for our discussion at this point, been created and are simply moving away from each other and away from a once very, very congested and incredible compact center. Generally, under this arrangement, the galaxies are also somewhat uniformly spaced throughout the universe. What irregularities do exist do not negate this contention. The point is, no matter how one slices the universe we seem to be at the center of things. Have we retreated to the early church doctrine that says we have to be at the center even though the church and science have arrived at a similar conclusion for different reasons?

Although I will shortly argue for a universe that is not expanding, Figure 2.2 would seem to be a more reasonable and non-coincidental arrangement. The small circle with the "X" would represent the observable universe within the range of our telescopes and would be populated by billions of galaxies (see photo on cover of the book again). The area outside the circle marked "X" is beyond our telescopic capabilities and would contain *trillions upon trillions* of galaxies, or an infinite number, despite the fact that our figure has only ten! Isn't it more likely that we are only seeing a tiny fraction of our universe? It would seem an unlikely situation that the limit of our detection is synonymous with the actual limits of the universe (if there are limits). This hardly makes us located at the center of the universe. In this case there could be no center.

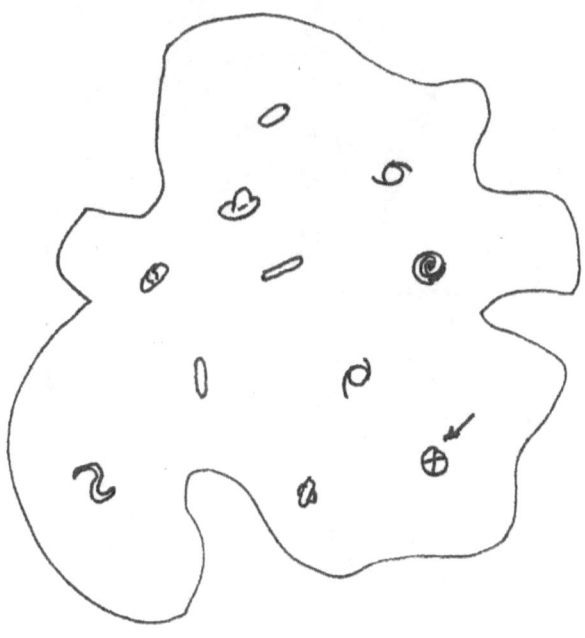

Figure 2.2

To make figure 2.2 more accurate the irregular outer line representing the edge of the universe should be removed to represent a universe without boundaries. This is a model for which I will adjudicate in a few moments.

Scientists tell us that we can see galaxies in all directions out to a distance of about 14 billion light years. In this figure 2.3 below (imagine it as a three dimensional sphere rather than a two dimensional circle), C and D represent the "edge" of the universe in all directions. We are located at A or B. If our position is B, it is 14 billion light years to the C-D edge of the sphere. We are then at some point "off center of the universe. If we are not at the center, then we are at location at some "off center" place such as A which is more likely. Then the interesting question is, "Which distance is 14 billion light years: AC or AD?" And how does this square with current cosmological theories?

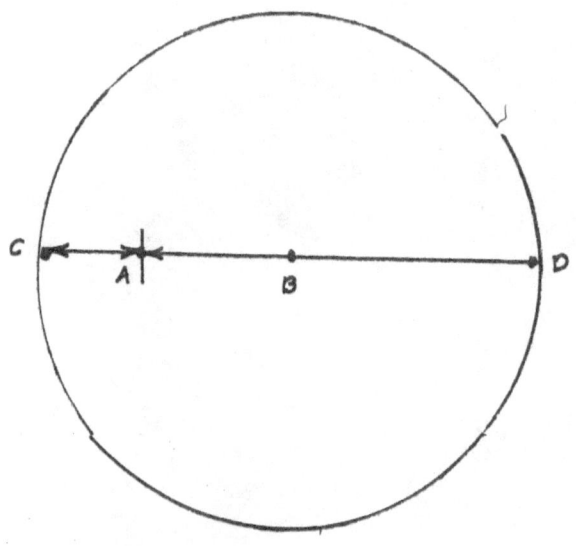

Figure 2.3

The answer, in current popular theories, is that there is no center nor "edge," but one great "curved" universe where no matter what direction we head, traveling at the speed of light, we will return to our original starting place after many billions of light years.

In this model, the universe has no edge, is curved and is expanding into a space that is not a space. This is about as sensible as the well known 18th Century philosopher, Bishop George Berkeley's denial of the existence of all matter.

To summarize this concern, no matter what scenario or model used to describe our universe, it hardly seems likely that we would be at the center of it all. It is certainly improbable that we have finally arrived at a technological capability to observe the galaxies that are the most distant from us but no farther. Thus there is no center. This philosophical perspective seems to insinuate itself into most scientific views about the organization of the universe. Perhaps in a layman's perspective, for every creature in the universe, no matter where that creature finds itself, that is the center of the universe.

I have mentioned the concept of the universe being like a balloon being blown up. The theory is that the universe is expanding, but that it is not expanding into empty space. I suggest that the universe is infinite with galaxies distributed throughout infinite space and thus not "expanding" (we will return to this later).

We can now deal with "empty space." To begin with we have one of those conundrums that hurts our brains to contemplate. When we consider our two rational options: either space ends somewhere (a belief near our comfort zone) or it continues infinitely. The former seems more logical until we realize that if it ends somewhere, what do we find at that point? A wall, a barrier or boundary of some kind; then what lies beyond that barrier? This is where the brain overload kicks in. If there is some indicator where space ends, wouldn't there be something there? Even if it is "nothing," to our way of thinking, "nothing" is "something."

The model where the universe continues on forever in every direction and never ends also staggers our comprehension. Our intellect wants to tell us it can't go on indefinitely -- it has to end somewhere! This takes us back to the first notion that there has to be a barrier of some kind at the end of the universe. Both possibilities seem impossible as far as our comprehension limits our imaginations.

Is there a third option to consider that may not be too palatable either. This theory is the most current among scientists today. They tell us that there is nothing – literally "nothing" - outside this expanding, balloon-like universe. The universe is not expanding "into" empty space because only the universe exists. It expands and only as it expands is there anything to exist. Empty space or "nothingness" just doesn't exist. This current theory is even more irrational than the previous two. Either there is an end to space or else space is infinite – both seem like impossible alternatives. I sense that the current theory is a strained attempt to deal with the previous theories because astronomy finds the first two incomprehensible. This third option makes no more sense than the first two, or even less. I would argue for the old fashion traditional idea that the universe or space is infinite in all directions -- the "steady state theory." This would comply with our axiom # 3: *The least complicated theory is usually the right one.*

One element of this third and more current theory is that space is curved.

If I understand this popular notion (and I'm sure there is a strong possibility that I don't), space itself, and not just elements of space, such as light waves and material things, are bent over large distances. The traditional view of space is that it is like an empty room where everything is vacuumed out – all the dust, air, the light photons, etc. There is nothing there except empty space or perfect vacuum. (And of course the vacuum cleaner.) Scientists suggest that even that "space" can be curved. Our view, if you are with me on this, is that there is nothing there to accept curvature. You can't smell it, see it, hear it (except cosmic background radiation left over from the big bang, which I question), or make any kind of contact with it. There is nothing to twist or bend. I would contend that the only things bendable or curable in a perfect vacuum would be light waves. The "nothingness" of the empty room cannot be affected by anything. "Nothing" cannot be bent or curved.

The "nothing" inside the expanding universe can curve according to scientists as contrasted to the "nothing" that exists outside the universe. We're told by these scientists that nothing really exists and there is no "outside" the universe.

The curvature of space seems to be a necessary corollary to the "expanding balloon" nature of the universe. I personally feel the idea of a universe constructed out of curved space is a feeble attempt by astronomers (whom I do like and admire) to cover up an embarrassing model of the universe that fails to make sense.

Some cosmologists have gone so far as to suggest there is no such thing as gravity. Instead the curvature of space/time is responsible for the movements of the bodies in space. For example, most of us think that without gravity

Cosmology and Theology

the planets would not be held in orbit around the sun, but would fly off into space in a straight trajectory. The curvature of space theorists say the planets would continue in their elliptical travel around the sun because the curvature of space holds them. This "gravityless" theory is almost totally ignored in most astronomy articles which I consider a suitable response to a far out theory. I will retain my belief in gravity.

I should pause and offer a disclaimer at this point that I would urge you to insert at various places throughout this material. Just because scientists believe they have proof negating my suggestions, I am not deterred. Eventually they will discover additional evidence that their earlier proof was wrong, and I will be vindicated. That could be a fifth axiom!

Let's pursue this curvature of space a little further.

The nonsense of curved space concepts is reflected in this interesting and embarrassing little drama. A contemporary of Plato's, one Archytas, was appropriately arguing for an infinite universe with no boundaries. The vehicle he used for illustration involved a person standing on the edge of the universe. That person sticks her walking stick out over the edge of the universe, then steps out by the end of the stick. She then proceeds to elevate and point her waking stick in that same direction over and over again to infinity. That is not a very ingenious proof of an unbounded universe. But it is far better than the rebuttal offered by a current day "reputable" cosmologist who argued against the illustration offered by Archytas and in support of the recent popular curved universe theory. This modern cosmologist will remain unnamed to avoid embarrassment. His refutation of Archytas's little play runs like this: The woman could stand on any spot on Earth and go through her little pantomime with the pointing stick and end up circling the Earth only to return back to where she started. Here is the cosmologist's clincher: his rebuttal involving the Earth, continues, "if Archytas's conclusion were correct, the Earth would extend infinitely in all directions. But it does not. It is a sphere." I think he misses Archytas's point.

Another cosmologist suggests that traveling in a "straight" line off into the universe and returning to the same spot eventually by way of a great loop is no more of a wonder than going around the Earth and returning to where you started. Somebody has gone over the edge! If we can't come up with better reasoning to support a curved universe, let's just drop it. I really believe scientists are just as bewildered as we are by the true nature of a universe that is infinite and are seeking a more acceptable model.

Probably the idea that space is curved stems from the fact that light is bent, as are electromagnetic waves and magnetic fields, by gravity. Plus as I mentioned above, curved space might be necessary to complete the expanding "balloon-like" nature of this modern theory.

One companion piece to modern theories of the universe is the suggestion that there are many universes – which is way beyond my grasp. Because this notion goes way over my head doesn't make it true. We are often impressed with the complicated and those ideas that seem beyond our intellectual capacity. Remember our third axiom: It is usually the more simple and least complicated version that is finally found to be the actual explanation.

Let us eliminate any multiple universe theory. Only one universe exists. However we can pause a moment to consider an interesting, albeit radical, possibility for multiple universes that we have all entertained at some time or another.

Noticing the similarity between galaxies and atoms we have all postulated the idea that our galaxies are in fact atoms of a much larger universe which is beyond our capacity to see. The only analogy I can think of is the novelty of nestled boxes where each box is smaller than the previous one and contains an even smaller one that also contains yet a smaller one. Each increment size fits inside the next bigger box. Such "stacked" universes could be simply all one universe where each atom is a smaller universe that is a part of the total universe that is stair-stepped. The sizes of these "stacked" worlds would be of an order of a speck of dust compared to a galaxy or incrementally greater in contrast – a juxtaposition of astronomically size differences with which astronomy is not unfamiliar. We can dismiss this model by simply asserting that an atom is categorically not a tiny galaxy.

Another possibility of a radical additional universe for astronomers to work with if they haven't already arrived at such a possibility would be a universe that is out of sync with the time of our current universe. Such a second universe would be in time slightly ahead or slightly behind our universe. See Figure 2.4. I don't agree with such a bizarre idea, but cosmologist like to play with unique ideas.

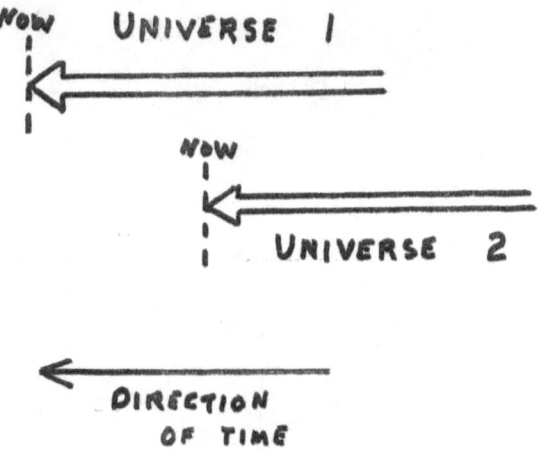

Figure 2.4

Either as a separate component or as a part of the many universe theme, we have the suggestion that there are multiple dimensions to the universe. We are all familiar with the usual dimensions we work with regularly. On a flat piece of paper we draw a line from left to right and a second line from top to bottom of the paper intersecting the first line. This gives us two dimensions. If we could draw another line that could extend from the surface of the paper plane out towards us or away from us behind the paper, that would be a third dimension. Or to use a cube (figure 2.5) where lines A, B and C, or edges of the cube, are all at right angles from one another, we have three dimensions. We are all aware of this and also with the idea that scientists often make time the fourth dimension. This is the world you and I live in. Cosmologists offer a universe where, when they refer to multi-dimensions, they mean five, twelve, sixty or any handy number of dimensions. Such a concept, because I cannot understand it, seems far fetched and meaningless.

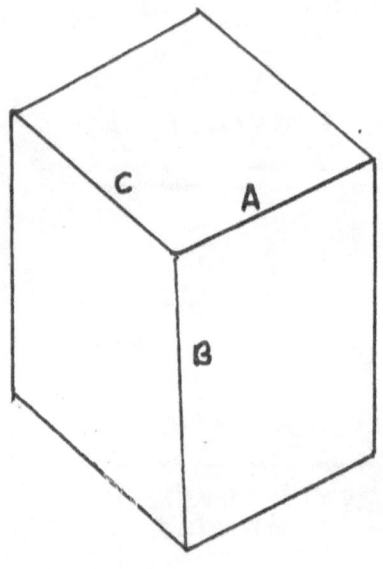

Figure 2.5

I too will offer a far fetched and perhaps meaningless theory. The universe has no dimensions! A dimensional system is an artificial construct that exists in the abstract only as far as the universe is concerned. It is a form of grid that we impose on space, whether it be the measuring of a room or using a map to navigate. Its purpose is to locate points in space in relationship to each other. In reality dimensions do not exist.

Because we can visualize a unicorn doesn't mean it exists outside of our imagination. Creating a mental picture of dimensions or curved space doesn't make it so. Proving an idea with mathematics, for example, doesn't make it exist. Is the system of mathematics another abstract construct artificially imposed on nature? Or is mathematics an integral component of the physical laws of nature (ignoring until the next section the existence of moral laws)? I would lean towards the former rather than the latter question suggesting that mathematics is not part of nature. It is probably a system of incremental ideas. The variety of measurements we use would seem to confirm this. For example, we can use our system of inches, feet, yards, miles, etc. or the metric system consisting of millimeters, centimeters, meters, kilometers, etc. But not together. The two systems work independently of each other and are not necessary. An infinite number of substitute systems can be developed comparable to what we currently use. Such is the artificial nature of the numbers we use for our theories and equations. Such systems do not exist in nature.

Examining the natural world we find many seemingly wonderful calculations that appear to belie what is actually happening. We don't believe that bees know much about geometry as they direct information to other bees concerning the location of resources for honey making in terms of directions and distances. Usually we chalk it up to instinct – something we don't understand very well. Mathematics serves us well as we send a space probe to rendezvous with another planet. Yet a dog's grasp of the calculations needed to run down a Frisbee are rudimentary at best despite the strong similarities between the two tasks. Throw into the mix the baseball player tracking down a fly in the outfield or a tight-end running under an arcing football. It's all done without imposing mathematics on the problem even though we would like to believe geometry is evolved. At least we know it isn't part of the equation for the baseball or football players. In the case of the dog we can't be sure.

Grizzly bears stand on top of a waterfall to make expert catches of salmon, who in turn are doing their own thing also without the aid of mathematics. Geese fly in "V" formations, to reduce wind resistance without knowledge of aerodynamics. No matter how much we try to impose mathematics on the situation, the goose would say that is only an artificial construct and not necessary.

The point being that we create artificial increments to measure and calculate, and more importantly. to "prove" our theories of the universe.

All this in order to say my theories or ideas no doubt will fail the test of proof by mathematics. An example of turning the tables would be my questioning the mathematical efficacy of claiming we need "so much material in any given space" to make the universe operate or function as it does. Or certain speeds need to be maintained to explain certain possibilities. I would question statements that galaxies must possess certain amounts of material to function in certain ways. Do we have that adequate of a grasp on the physical laws of our universe yet?

Is it now time to unveil the true or actual structure of the universe? Utilizing the third axiom once more, one could argue that the simplest structure for the universe would be one where it is not expanding or contracting or curved or multiple, etc., but is rather static. I am attracted to a theory of the universe that is infinite in size and composed of an endless number of galaxies. The galaxies appear to be "swirling" gently through space (in actuality, they are moving at incredible speeds) in random motion like fish in an aquarium. They are pulled in different directions by gravity, sometimes being pulled apart by attraction from a larger galaxy, sometimes colliding and merging. This is much like the swirling motion of our "local cluster" of a hand full of galaxies in our neighborhood.

Immediately the red flag goes up as everyone says this contradicts the evidence of the "red shift!" Let's explain the phenomenon of the Doppler effect for any readers who may not be familiar with it. Use figure 2.6 to follow this discussion.

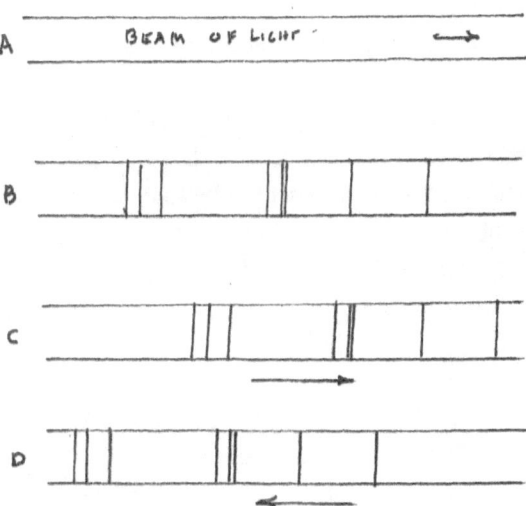

Figure 2.6

To begin, the spectrum of light reflected from an element (iron, helium, radium, etc.) resembles the ubiquitous bar code found adorning almost all the merchandise you purchase today.

Line A in figure 2.6 represents a beam of light that has passed through a prism and been split into a spectrum like a rainbow. At one end of the spectrum is blue followed by green and yellow and orange, with red at the other end. What our eyes see as white light actually contains all these colors. The prism spreads the different wave lengths of light into the color spectrum. Blue light has a shorter wave length than red light.

Line B shows the spectrum when some element – iron, sodium, helium, gold, neon, carbon and so forth – has absorbed some of the frequencies or wave lengths when light is reflected off of it. Light reflected from each element will have a different set of dark lines in the spectrum – each pattern unique to the particular element. It is like the element's finger prints, if you will. Helium was discovered in this way in the sun before it was discovered here on Earth. The dark absorption lines in part B are some element when the source, the sun or another star, is stationary or not moving in relationship to the observer or the spectrum.

Now put the star, or source of light, in motion in relation to the observer or spectrum and the lines of the element's "finger prints" will shift towards the red end of the spectrum (represented by line C) if the source is moving away from the observer. The "finger print" will shift towards the blue end (represented by line D) if the source is moving towards the observer. Physicists often use the example of a train to illustrate the phenomenon with sound waves rather than light waves. Stand at a railroad crossing and listen to an approaching train. As it nears you the sound is higher pitched because each position of the origin of the sound is closer than the previous one, making the sound waves closer together. It would appear, if you were diagramming it, as an accordion box in its compressed stage with the folds closer together. As the train passes you and moves away (assuming you were not run over) the sound goes lower in pitch because each position of the origin of sound is farther apart and the sound waves are farther apart. Imagine the accordion extended and the folds are farther apart. The faster the source is moving, the greater the shift will be. This is how astronomers know whether a star or galaxy is moving towards or away from us.

The problem is, with the exception of the local or nearest galaxies, almost everything seems to be moving away from us and the more distant the galaxy, the faster it is moving in relationship to us! The message this sends is that the universe is expanding in all directions away from us. Should there be an exception to the expanding universe where everything is moving away from us? The closest galaxy, Andromeda, which is only a little over 2 million light years away from us, is not moving away which should be the case. It is instead moving towards us at about 260 miles per second.

Extrapolating backwards from the current situation, assuming that the galaxies are moving away from us, it would support the popular theory that billions of years ago it all started out from one singularity or at one place or point of origin. The explanation arrived at from this theory is the popular notion of the "Big Bang." There was nothing. Then matter and the universe (its beginnings anyway) burst on the scene from nothing and began expanding until it has reached the current dimensions today. Where it all came from to pass through the window of the big bang moment into existence is not explained. No one has an answer to that question, so don't ask it. The universe just appeared as a singularity which then expanded.

That leaves us with two possibilities (at least two simple ones). It all started with a big bang briefly outlined in the previous paragraph. Or as another simple option, the universe had no beginning. It has always existed very much in its present shape, with constant "minor" changes in structure. (I say "minor," meaning that overall it has always looked much as it does today even though the galaxies are making "major" changes in shape and in relationship to each other.)

You have before you two theories regarding the origins of our universe: the Big Bang theory and "it has always existed" theory, known as the the Steady State. Either or both theories leave us with the unpleasant stretching of our comfort level. Neither one seems plausible or reasonable! Which seems most likely? Everything came from nothing is totally impossible! But then the idea that it is eternal in both directions, forwards and backwards - it has always existed and always will – in turn is also totally impossible! The limits of our imaginations make both theories impossible! Neither one can be! The same dilemma presents itself concerning the existence of the universe at all. Think about it. The universe can't possibility exist! And yet, it can not possibly not exist! Neither seems reasonable and yet it exists. At least one of the two impossibilities is real: the universe does exist (unless you hold with Bishop Berkeley's theory that nothing is real – which I believe contradicts itself).

Remembering the third axiom once more, the simplest of the two theories – the universe has always existed in its present configuration – seems most reasonable. I want to accept this possibility and make a case for it. Which makes for an interesting turn of events. The Christian church has argued for a universe that was "created" (had a beginning) and a universe in which we are at the center. Science has, for the most part, come around to the church's position on both of these points. I would like to argue against this position on which science and the church seem to be in agreement. It seems more reasonable to accept the concept of a universe that has existed without beginning – and that we are not at the center of our universe since infinity can have no center.

In order to support the suggestion that the universe is not expanding it will be necessary to deal with the main evidence that suggests the universe is expanding – the Doppler effect. The fact that, with the exception of a few local galaxies close by, all galaxies are moving away from us (the red shift) at speeds that increase with the respective distance from us sends us the message the universe is expanding.

Before engaging the red shift/expanding universe problem, we might pause to examine other problems. According to observations based on the red shift of their received light, the galaxies at the far limits of our current visual capabilities are moving at near the speed of light. One problem with this is that any galaxies "further out" by extrapolation are moving faster than the speed of light! Since we are told nothing can move faster than the speed of light, this would be impossible. To circumvent this problem astronomers tell us that the galaxies are not moving, but rather the universe is expanding! The galaxies are just moving (ops, the wrong word) – being carried along with it. I like to think of myself as a progressive thinker, however, to me any galaxies beyond that "edge" are "moving" and are not just being carried along at the speed of light.

A related problem is that most astronomers, I believe, seem to agree that the farthest limit of our visual ability at the moment (where objects are proceeding at near the speed of light) is the edge of the universe. In all directions! Which places us at the center of the universe again. This edge is, we are told, in the neighborhood of 14 billion light years away (current best estimate is 13.7 billion light years). For the edge of the universe to be at the limit of what we can see in all directions (14 billion light years away) makes for one wild coincidence, as mentioned earlier. It coincides with the limit of our current visual capabilities and is the same in every direction. I use the term "visual" when, in fact, some of our observations are not visual, rather, they are radio emissions or in the radio portion of the spectrum.

Referring back to the suggestion that the red shift evidence makes it seem to refute the theory that the universe is not expanding. Instead it is static and analogous to dust particles swirling slowly in all directions while suspended in a ray of sunlight.

Is there any other explanation besides a rapidly expanding universe to account for the red shift? One possible cause for the red shift would be that whatever constitutes light rays or waves is slowed down over such vast distances by exhaustion or fatigue. We know that gravity pulls on or affects light, and the gravity over billions of light years distance could provide a drag or exert a force to account for the red shift. The additional influence of the great amount of "dark matter" that astronomers tell us exists, if it does actually exist, would contribute also to a slowing of light. Light loses it's magnitude or illumination power over distances – a proven fact. Couldn't the speed it travels be diminished? Scientists say that the speed of light is constant. But is this true over great distances when affected by gravitational pull of many, many intervening galaxies plus a natural weakening? If there were no diminishing of the speed of light rays over distances, wouldn't that eventually be tantamount to perpetual motion? If light over immense distances slows down as it weakens also in illumination, it would seem natural as the two properties are part and parcel of each other. In addition, this would be consistent with the red shift increasing with the distance.

You may recall I mentioned above that our closest neighbor, the Andromeda Galaxy, is moving towards us at about 260 miles per second. Thus there is no red shift. Perhaps the cause of the red shift is, as I suggested, due to a light fatigue over great distances which would account for the anomaly of our closest neighbor not being affected to the degree our other more distant galaxy neighbors are. It seems to support my theory rather than the current idea that everything is moving away from us. An expanding universe should carry everything with it – at least the kind of expansion astronomers suggest where the galaxies are not moving away, but space with all of its contents is expanding. In that case would there even be a red shift?!

Remember that each theory intending to describe our universe is not without its unexplained problems. Such enigmas don't necessarily disqualify a theory. They only provide another mystery yet to be solved. And I should repeat every so often in this section concerning our exploration of our universe, that scientists, astronomers and physicists have serious difficulties with measurements on the infinitesimally small and the immensely large scales of our universe. This accounts for the weird effects of quantum physics as well as the trouble with dimensions on the order of light years.

Scientists are always calculating the amount of physical matter (including strange matter) that is necessary to be consistent with the explanation of an expanding universe. Too much matter or too little matter will determine whether the universe will expand forever or eventually slow down and retract. A universe I have just described that has no end or limits is neither expanding nor contracting. The galaxies are simply swirling through space occasionally colliding and merging or pulling one another apart to form new configurations. They are simply drifting randomly. No particular or specific amount of matter is necessary or required to stabilize, balance or energize a static universe where galaxies are spaced in no special arrangement throughout infinity. According to axiom three, the more simple version or description of the universe will likely be the closest to the actual situation.

Some astronomers refute the theory of an infinite universe that is made up of endless galaxies by suggesting that such an arrangement would cause us to observe a galaxy at every point in the sky. No matter where we look there would be a star or galaxy – every singular point where we direct our vision. Reality negates such an idea. Simply observe the cover of this book once more. We cannot see objects, be they stars or galaxies, that are too far and thus too dim for perception! The smallest or furtherest galaxies in the cover photo can only be detected by telescopes after several days of exposure! And then only with radio telescopes, not visually. Objects a little more distant are undetectable, even with the aid of our best telescopes. So much for the argument that in an infinite universe we would be observing light coming from every point of space. It's not going to happen.

In a related thought, at least some astronomers wonder why the light from the many stars and galaxies doesn't light up the sky at night like a supermarket parking lot. They think such a phenomenon should persist and wonder why it doesn't. My discussion regarding the red shift fatigue provides an explanation.

Cosmology and Theology

No matter what theory is offered, there are irrational elements because our minds cannot comfortably accept what seems impossible. That the universe is infinite is irrational. That everything – all matter – came from nothing is irrational. That the universe has an edge or end to it beyond which there is nothing, not even empty space, is irrational.

Let us return to an earlier discussion in this chapter, the idea that space is curved. It is one of the more difficult ideas to swallow. I suppose the idea is put forward to allay skepticism concerning a universe expanding into nothing. Figure 2.7 is a simple diagram to illustrate the confused nature of a space that is curved.

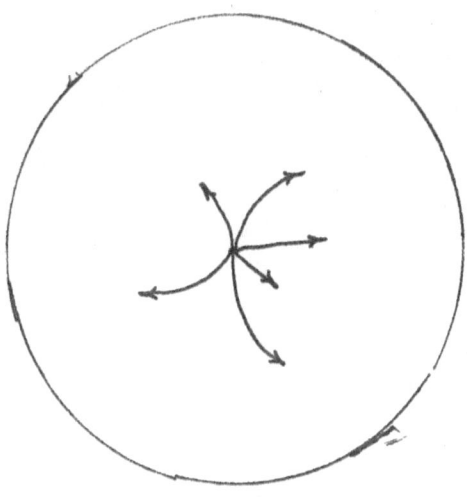

Figure 2.7

Imagine the circle (which represents the universe) is three dimensional which makes it a sphere. How do we diagram curved space? Would the curved arrows in figure 2.7 be adequate to express such a possibility? The galaxies appear to be moving away from us in what may be presumed to be somewhat straight line trajectories. If the roughly 14 billion year distance as the parameter or radius of our universe is accurate, are the galaxies in that vicinity or region beginning to make a sharp turn? No, not if the universe is expanding and carrying the galaxies with it. Where is the curvature? And more critically which direction does the universe curve or bend in such a spherical configuration?

In a universe such as I propose that goes on infinitely with no edges or ends, there is only space. There is no "non-space" outside the universe. There is no "outside." Consequently there is no room or need for curvature of space. The only things that bend or curve are light rays or waves which are bent by gravitational pull of matter. This curvature is slight compared with the great distances covered and is entirely random in direction. Such bending of light would be irregular, at the mercy of whichever galaxy or galaxies are the closest and largest. The light rays would bend first one direction and then another direction as shown in figure 2.8.

Cosmology and Theology

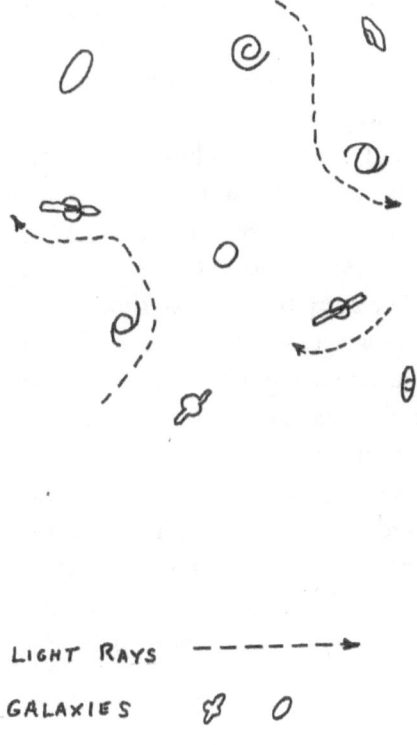

Figure 2.8

Regardless of which theory of the universe you favor, most of us probably agree that the farther we look out in space the further back in time we are observing. Three possibilities present themselves.

First, if matter, and thus galaxies, are being created as time moves along, at some point of origin this would be analogous to lava bubbling up from the ocean floor to connect existing tectonic plates making up the Earth's crust, or to a spring of water bubbling out of the ground, then matter is constantly being created to feed the material needs of an ever expanding universe. Creation would be continuous, but where would that point of origin be?

Second, if all matter was created at once with the Big Bang and is only spreading out farther apart in space, then all galaxies would be the same age. However, because we would be observing galaxies at different stages of evolution the farther away they are from us, there should be a continuum or progressive incremental change in evolutionary appearance in conjunction with distance from us. Andromeda, the closest large galaxy to us, would appear much like our Milky Way (our galaxy) as it would be similar to our stage in evolution. And it does. However, in each succeeding group of galaxies as we move away from us, we should see a somewhat smooth transition from current age to what galaxies would look like earlier in their history, much as we see regular incremental changes in the evolution evidence in fossils that paleontologists uncover for us. The difficulty of finding a smooth history of change in fossils is exacerbated by the tremendous upheavals in geologic processes as well as the slim odds of fossils being preserved in the first place. However unlike with fossils, the array of galaxies in a line of progression stretching out into space is for the most part complete and should display a nice evolutionary arrangement. But as I read the scientific literature, I don't believe such a clear transition is observed.

Third is my suggestion of an infinitely large and static universe which makes a continuum of galactic evolution a moot point. The universe is ageless. It is not expanding. Galaxies do not evolve in regular transition. They are at different stages of development in a cyclical pattern. Such a configuration, I think, conforms to what is actually observed through our telescopes.

Examine the cover photo of this book once more. The many galaxies pictured there are so far away that this photo required ten days of exposures to accumulate the light from the most distant objects. We are told that the most distant galaxies (and consequently the farthest limits of our perception) are in the neighborhood of 14 billion light years distant. This could be off by some two, three, or more billions of light years either way due to the great difficulty of actually measuring distances of such vast magnitude. The light we perceive coming from such distant objects left that region billions of years before our planet Earth was formed and is just now arriving here! Astronomers assume the faintest and most distant objects are at the edge of our universe and that there are none farther away.

Cosmology and Theology

This in itself would seem to me to be a shaky conclusion. Nevertheless, for the sake of making a point, let us accept the hypothesis that these objects are at the known and actual edge of our universe and that they are 14 billion light years away. How can we say this makes our universe 14 billion years old?

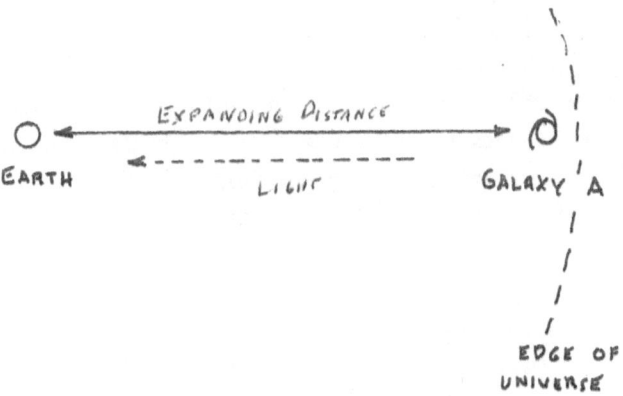

Figure 2.9

If I am wrong and the prevailing theory of the expanding universe is right, consider the following scenario. While studying figure 2.9 and assuming the popular theory of an expanding universe, note the galaxy at point "A" that is near the edge of the universe and thus roughly 14 billion light years away from Earth. What about the great amount of time it took the universe to expand to this great size? If it takes light from these distant objects fourteen billion years to reach us from there, wouldn't it take an equal or greater amount of time for those objects to reach that distance after the Big Bang, assuming there was a Big Bang and the model of an expanding universe is accurate? Consider that these galaxies are moving slower than the speed of light. So while it took light from galaxy "A" 14 billion years to travel the distance from "A" to Earth, we need an additional 14 billion years plus many more for the galaxy to get from Earth to its position at point "A". The assumption being all matter expanded originally from, a singular point - the big bang. By my math this makes the universe more than 28 billion years old. Only because I am unable to comprehend the astronomer's math.

Note: 14 billion years for light to reach Earth from point "A" plus an unknown number of years for "A" to reach the edge of the universe ("A's" location in figure 2.9). Where is galaxy "A" now?

Take the model of the universe I have proposed and you needn't bother with distances and ages. I like the idea that it is infinite in space and time! Remember the third axiom: Usually the simplest explanation is the most accurate one. My theory fits the third axiom.

Perhaps we should re-examine the first axiom at this point: The universe makes sense. History is a record of phenomenon that seem impossible or ridiculous until more is revealed and understood about the subject. At that point we come to understand the principles involved and find that it is not as strange or impossible as it first seemed. What appeared to be a mystery, upon further investigation becomes logical. It now has a reasonable explanation. This can be illustrated with airplanes and flight. At least over 100 years ago the idea that people would ever have machines that could fly was out of the question, though they were flying in hot air balloons a hundred and fifty or more years ago. Suggesting such an idea would have only brought you ridicule. It was considered impossible to get a large body or a human being off the ground for extended flight. Now that we know the logical principles involved it is perfectly reasonable.

Much about the theories concerning our universe appears to be illogical at the moment. Our first axiom tells us that, if we had access to more information, much of what seems strange or impossible to us presently would become comprehensible and reasonable. Unfortunately, or fortunately, the universe is so big and so wonderful there is much we will never unravel. There will always be mysteries beyond our reach!

The first axiom that the universe is logical and reasonable and makes sense, applies to all physical laws. As we will see in the next two chapters of this book, it applies to moral laws as well. In addition, the fourth axiom claims such moral and physical laws are ubiquitous - constant and consistent throughout the universe.

To flesh out the nature of our universe, we should consider a few other topics: Life elsewhere in our universe, creative design versus evolution, and the absolute or relative nature of size.

Does life and intelligent life exist elsewhere other than our planet Earth? You might well ask does intelligent life exist anywhere at all?

Many people are intrigued with the possibility that there are other places in our universe that have life of some form: 1. Are there other planets with life? 2. What is that life like? 3. Will we be able to communicate with that life if it is intelligent and is technologically advanced as we are? The answers to questions one and two are easy. Question three is problematic.

The answer to the first question is a resounding yes. There are some 200 or 400 billion stars in our galaxy – the Milky Way. We have discovered planets circling some stars already, and the number of such discoveries is increasing rapidly. Extrapolating from the small sample already detected, astronomers are now estimating about 100 billion planets in our own galaxy alone. And there are billions of galaxies in the universe, each with comparable numbers of stars. Maybe an infinite number of galaxies if the universe is infinite. You might well ask, "Is there furniture in my neighbor's house since I have furniture in my house?" Yes. Although not all houses have furniture in them. Not all stars necessarily have planets. But a good guess is that most do. Not all planets have some form of life. Most probably do not. Look at our solar system. Neptune, Saturn, Jupiter, Venus and the other planets are lifeless. One or two of the solar system's moons offer a slight hope of harboring life, and that hope does not include creatures that drive cars and watch TV.

Yet if one star out of every billion has a planet with life and only one of those planets out of every billion has life comparable with human intelligence, there would still be a vast number of "Earths" with intelligent life. An intelligent and reasonable guess is that every galaxy that has evolved to a certain point has many planets hospitable enough to have life in varying stages of development. We have answered question two along with question one. There is probably all kinds of life in the universe – even some like you and me.

Now concerning communication, the distances are so unbelievably great between stars and thus between planets having potential life, that contact at best would have to be by way of radio signals. The nearest star to our sun is about three or four and a half light years away (perhaps more than 40 trillion miles). If we could travel at the speed of light, something we can't even begin to approach, it would take at least nine years for a round trip. With our current technology, a more realistic estimate of the length of time to make a trip to our nearest neighbor star would be more like over 50,000 years. Most of the stars we see in the night sky are hundreds and thousands of light years away. Any radio signal from a "near-by" star's companion planet could take many years to reach us, and our reply would take the same amount of time to return. Not a very lively conversation! We have yet to detect any such radio messages, but not for lack of trying. In fact we have been engaged in exhaustive searches with no success yet.

If we were alone in the universe and Earth was the only place on which life was found, it would seem that the rest of the universe (which is incredibly big and measured in terms of billions of light years) is an amazing waste! There would be no creature or consciousness to appreciate it all. We could well ask what is the purpose of the rest of the universe if no one's out there to enjoy it? Attributing a purpose to the universe is a big assumption and you could respond with, "You are guilty of imposing our limited human values on the universe and that is inappropriate." My defense is the second axiom. We have nothing at all if we do not use our own value systems. Given a choice of there being, or not being, a reason for the existence of our universe, we should assume there is a reason. Remember our agreement to accept the first axiom: The universe makes sense. Besides I need a purpose for it all to validate chapters three and four of this book.

I believe the odds are incredibly good that we are not unique in our universe. Countless planets must exist sustaining forms of life from early stages to levels no doubt beyond our own. We are neither unique nor alone in the universe. We just may never prove it.

It would seem to be a natural leap from the topic of life elsewhere in the universe to discussion concerning our next topic: evolution and intelligent design.

Scientists tell us there are critical factors exactly within certain windows for life to exist on our planet. These factors would include gravity, the Earth's size, our distance from the sun, the existence and balance of chemical relationships, and other environmental considerations. For example, the attraction that binds atomic nuclei together must be an extremely fine-tuned amount or matter itself could not exist. One tiny shift or change, plus or minus, and poof, we have nothing. It sounds almost as if it had to be intelligently planned!

On the other hand, evolution is a given.... a proven fact! To say evolution is not true is tantamount to saying gravity does not exist. An evolutionary process is at the heart of almost all science. It is the same basic foundation for change. Take it away, and you have denuded astronomy, biology, botany, zoology, geology, paleontology, physiology, genetics, archeology, etc. For example, earlier we noted that astronomers tell us that each atom in your body (one tiny atom in your fingernail) was once in the interior of another star other than our sun. The star exploded as a nova and spewed material out into space only to coalesce millions of years later into another solar system where you live now! This may have happened more than once.

Evolution and the idea of creative design should not be adversaries in a court case, "evolution vs. creative design" (Scopes Trial, 1925). The two ideas are not mutually exclusive! It would not be stretching it to say they might complement each other.

The evidence for evolution born out by the different fields of science mentioned above, as well as others, defies refutation. On the other hand, any casual examination of our world which includes the vast universe that most persons are not as well acquainted with, which would include the beauty of the stars, galaxies, awesome planets and moons, the grandeur and indescribable and colorful beauty of nebulae stretching over distances measured in light years, etc. In addition to those wonders, we live in close connection with our earthly environment. One gets an overwhelming impression, right or wrong, it must have been intelligently designed.

Consider the marvel of our eye sight and our ability to perceive, and the computer brain with its wonderful functions. A butterfly that migrates by instinct over thousands of miles of potentially vicious weather, a giant redwood tree growing from a tiny seed, the marvel of human reproduction that we describe as "a miracle."

There are far too many incredible things and processes too numerous to catalog. Such a list would have to include what I enjoyed in the Canadian rockies on a recent trip: Lake Louise with its backdrop of incredible mountains and the marvelous variety of beautiful flowers in a garden in Banff. This latter, you might say, was not part of creation, but was developed by human ingenuity in plant breeding. Such wonderful and gorgeous flowers did not exist originally, but were bred by us from less spectacular specimens. This does not negate the idea we are examining. It only enhances the amazing process and existence of all creation as the fact of our creative ability to propagate lovely flowers and their "potential" to be developed venerates our universe! Such potential for creativity by human creatures as witnessed by television, telephone, space probes, medical achievements, etc. all shout an orderliness and integrated wonder that is spectacular! Enough of the superlatives!

It all stretches far beyond our imaginations which, in themselves, are no mean creation! Could it be, with all it's awesome aspects, our universe was "planned" with evolution as the method of creation? Could it have occurred by happenstance? Is it all just a random accident? Such considerations leads us naturally into the next section of the book: is there a god and what would a god be like? Other items related to the subject of this section you want to know about would include "black holes" and "quasars." How do they fit into my description of the universe? I don't know.

As you know, a black hole may be a star that has grown so large that its immense gravity is too powerful to allow light rays to escape. The star's light is bent by the incredible gravity back into the black hole. If the light doesn't come away from the star we cannot see the star – a black hole. The presence of black holes probably does not attest to the veracity of either theory (an expanding or static universe) more than the other one. Black holes could be equally at home in either kind of universe. Nevertheless, I am not confident that black holes exist. If black holes lurked around in galaxies (or we're told at the center of galaxies), not today or tomorrow, but sooner or later, it would seem likely they would eventually eat up everything including the entire galaxy! Somehow this seems to violate the third axiom, simpler is better.

My explanation for what seems to be evidence for black holes would be that the mystery surrounding a black hole is too unfathomable to verify its existence. Our ability to measure anything on the large scale we are dealing with is shaky. Computer models of black holes do not impress me. The skeptic in me wants to say a computer model cannot reliably replicate what actually happens to giant or super giant stars at the end of their lifetime. I am partial to the idea that no star reaches black hole proportions. At best, an immense star's pressure reaches a point where it explodes as a super nova before reaching black hole dimensions. Current accepted theory of black holes postulates an insatiable appetite where a black hole swallows everything that comes its way with no known limits. This is the reversal of the Big Bang where everything explodes from a singularity. If a black hole existed, it may be crunching everything back down into a singularity.

Quasars are another mysterious celestial object that emit incredible amounts of radio waves and/or equally incredible quantities of light. Their power is one more display of the grandeur of the universe. Scientists are unable to explain their nature or what caused them. Thank heavens they are only found at vast distances from us. One in our vicinity could be very hazardous to our health. A good question is, Why all of them are so far away and none are close? Does their arrangement in space speak in favor of an expanding universe rather than a static one? For now they must remain a mystery.

As a philosophical interval, ponder the scale of our universe. Are the comparative sizes in the universe absolute or relative? Everything is measured from the perception of the observer. To an ant a marble is analogous to a large bolder for us. From the window of an airplane a car on the road below us is tiny. For the folks riding in that car, the car is full size and the plane is tiny. You say it is only perspective. However, remove everything from your presence or environment. There is just you and nothing else. What do you measure yourself by? Now are you any specific size or no particular size at all?

Consensus might affirm that size is absolute, but where would the evidence be? If size were relevant there would be no end to how infinitesimal matter can be just as there is no limit in size of an infinite universe. It is easier to imagine infinity on the large scale than it is to speculate on the "infinitely" small. Physicists have difficulty understanding quantum mechanics (matter and energy properties on the minuscule level) and its strange behavior. Evidence for an absolute size for everything might rest with the atomic or quantum level. The speed of light seems to tilt the argument toward the absolute over the relative. Gravity may or may not be a determining factor of evidence either way.

I get the impression that sometimes the theories astronomers espouse are much like what I would call the art gallery parable. A group of people are viewing a large painting. The painting is a bright red canvas with a black dot in the corner and a yellow diagonal line in the opposite corner. The viewers express skeptical opinions: "I could paint something like that." "I don't believe that's real art." "I think the artist is pulling the wool, as well as the canvas, over our eyes." Along comes a respected art critic who claims the painting is a fine piece of art, worthy of the $5000 price tag. The opinion of the art critic influences the thinking of the group because of her or his reputation. None of the viewers want to appear art illiterate. So now the consensus of the group is that the painting has artistic merit to those who are true connoisseurs of great art.

Do we often respond the same way concerning the far out theories of cosmology? - Theories of black holes, anti-matter, curvature of space, string theory, multiple universes, dark matter, a claim that the number of atoms in our visible universe adds up to ten to the eightieth power, multiple dimensions beyond four, and strange behavior of quantum mechanics must have merit since many professional cosmologists support the ideas.

I'm going to predict the "steady state" theory will make a come-back in the future to replace the current "Big Bang" theory that is now the predominate current astronomers' thinking.

It would seem logical that in an inflationary universe, there should be no colliding galaxies. Yet galaxies are colliding. The expanding universe should carry galaxies away from each other and eliminate any collisions

Cosmology and Theology

or physical contact. Secure a good picture book on galaxies and enjoy the distortions of colliding galaxies such as the "Two Mice" galaxies that appear to be merging. As you enjoy the exciting configurations brought about by colliding galaxies, think about this. Two galaxies can pass through one another without any star collisions of any of their billions of stars despite the gravitational distortion of each galaxy's structure.

I would like to close this chapter with two summaries. The first summary pulls together the central ideas in a brief statement using Figure 2.10 as a guide. The second summary briefly states ideas from this chapter that may help us define the material that applies to the existence of a deity that will be discussed in Chapter 3.

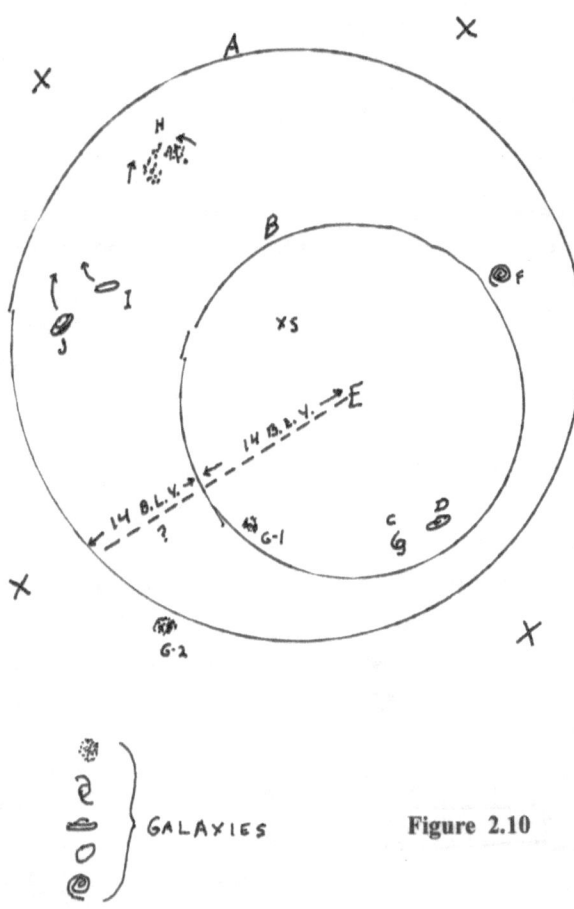

Figure 2.10

Legend to Figure 2.10

- E is Earth and Milky Way
- A is the "edge" of the universe (though there isn't any edge)
- B is the extent of the observable universe
- XS is the singularity or Big Bang (location unknown)
- "Cosmic speed limit" is 186,000 miles per second
- H is the site of colliding galaxies
- F is a galaxy outside observable limit
- Dashed line is a light ray
- C, D are galaxies furtherest out in our detectable field
- J, I are galaxies curving in space
- G1, G2 are two positions of a galaxy
- B.L.Y. Stands for billions of light years
- X is an area of nothing (not even empty space!)

* * *

Partial summary of difficult problems, questions, mystery, wonders, controversies, inconsistencies to ponder

- Astronomers think B is the limit or edge of the universe and by their description, we are the center of the universe which seems highly unlikely.
- Galaxy G1 is the location of a galaxy at the edge of our observable universe that has sent light beaming back to Earth over a fourteen B.L.Y. of time. G2 is the position of the same galaxy currently having traveled further out during the time it took for light to reach us from the G1 position and is beyond our detection. What would the curved space do to the path of G?
- Telescopes have revealed galaxies colliding. It seems there would be no collisions in the expanding universe that astronomers describe. They would all be diverging from each other.
- Astronomers talk like there is an edge to the expanding universe and at the same time, tell us there is no edge because space curves.
- In Figure 2.10 galaxies C and D are located about as far away as anything we are able to detect with our current technology. We

are told that all galaxies are moving away from us and the further away, the faster they move so that galaxies C and D are moving away and very close to the speed of light. Extrapolating from this information when C and D move further away, they will soon be traveling at the speed of light and finally even faster which is impossible. We are told that nothing can go faster than the speed of light because it is the cosmic speed limit.

- In the expanding universe current astronomy describes galaxies as not "moving" as they are instead being carried along with the expansion. Does the "carrying along" negate the concept of the red shift?
- It seems unreasonable that, considering the existence of countless billions of possible planets in the universe, that we could doubt the existence of a wide variety of life forms as well as unlimited stages of geological and biological evolution.
- That the universe exists, among other ideas, seems improbable. Yet we had better believe it!
- Our universe is incredibly astounding. At the speed at which light travels, it still takes billions of years for it to cross the universe.
- Is it infinite and how could that be?
- It is shrouded in mystery. We cannot grasp or explain very much concerning its construction.

No matter what or whose theory we examine, we are confronted with conditions that seem impossible to comprehend.

- It is filled with beauty from a daffodil to a nebula or a spiral galaxy.
- It is full of calamity and violence.
- It reflects profound evidence of a "creative design." Even with all the confusion of the conservative view attached to the phrase, "intelligent design," we may be able to salvage some significant meaning from the concept without doing irreparable damage to evolution.
- What can we take from an exploration of the universe that would become a basis for an understanding of God, if there is a god? What about the nature of the universe tells us there is, or is not, a god, and what God would be like?
- Remember the second axiom, the universe must make sense.

- Does death end everything and destroy all life or is there some kind of existence or experience of consciousness beyond death?

I will again restate my purpose – To impress you with the awesomeness that would suggest an intricate structure behind the universes creation. If nothing else, I hope this chapter generates more interest in our universe.

3

GOD

God can't exist! And God cannot not exist! Which ever is the actual situation, one has to be true and one false. Either or both seem impossible even in our wildest imagination. Just as we found it difficult to believe that the universe exists: it is not possible. But then it is not possible that it could not exist. Our emotions and intellect tell there has to be something. These thoughts are akin to two more impossible concepts: the universe can't continue on into infinity, and yet it cannot end. If it ends, what is beyond the end? Struggling with two contradicting ideas where, both seem impossible, yet one option has to be true, makes our brain hurt. Let's go with "God does exist" and work with that.

There is the universe; there is empty space; and there is matter. There are also ideas, truth, and knowledge. Almost all of us agree that some things exist. For the non-astronomer, nothing is "something." Something triggered our environment and everything else. There is SOMETHING! Actually we are all in agreement that something exists and that something was started by something. We are all trying to guess what that something is. If you take everything material out of the universe, and leave only empty nothingness because you have vacuumed every speck of dust out of empty space, nothing is left (except the vacuum cleaner, of course). Nothing is left and that nothing is something for most of us. Even if astronomers can't see or acknowledge "nothing."

Permit me a literary license to change a few words of a popular poem.

> "Yesterday I saw a universe that wasn't there. It wasn't there again today. Gee whiz I wish it would go away."

Part of the problem is that we need, not just a definition of God, but a mature concept that answers our observations of the universe and life.

Among the many concepts of God, there is the popular notion of God as the explanation of everything that is left over after science has revealed the way the universe functions in some new specific way that used to be one of God's domains. Step by step, in this scenario, God used to be responsible for everything until science loosened the connection between God and the issue at hand with an explanation that eliminated the need for God from the process or moved God one more step away in the chain of events. After science shows us the method of evolution as the origination of species, reveals the way stars and galaxies were formed, explains how life began with a lightening strike in a primordial soup (or whatever), and so forth, the need for God to be part of the picture seems to become more irrelevant.

Finally as a last refuge for God, God becomes relegated to the corner of the first cause no matter what else science has co-opted. Theology says to science, "You can explain everything except how it all got started in the first place. We have the ultimate answer! The Big Bang theory doesn't really answer how all that energy and material came from nowhere. And into what *space* or *territory* is all this physical material and energy expanding? We have the trump card; God did it! See, theology is not "totally insignificant after all." Everything seems to be working out well as theology has a strangle hold on the ultimate explanation because science can not really explain how everything came from nothing; theology has the answer, God did it. That is until that little child in the first grade Sunday school class asked, "But who created God?"

As I said in the beginning the purpose of this book is to explore the nature of creation and from what we learn about creation to explore the nature of God. We need to speculate, using what we believe about the nature of God to teach us about the nature of immortality or what happens to us after death.

Our task is to explore the nature of God and from logic and reasoning, what must be the nature of our fate after death.

You may well ask that since the majority of the handful of people who read this book are Christians, why don't we simply go to the Bible and read what it has to say about God and look to the words of Jesus for knowledge about an afterlife. In doing that we would miss the enjoyment of our own personal journey in an exercise of deduction. We can stretch our imaginations by examining what our universe reveals for us and hold the Biblical answers in abeyance. Once we have exhausted the struggle with the logical possibilities ferreted out through our personal search and ruminating, then we can go to the scriptures. However, you know you have already done that. Just don't let your perceived Biblical notions prejudice your enjoyment of your search for answers elsewhere. It is as if your neighbor asks why you are wasting your time in your vegetable garden when you might simply walk down to the store and buy some canned goods or fresh produce. Your neighbor may not understand that you enjoy planting, cultivating, harvesting and how that could be more meaningful for you.

If you feel compelled to wear another label, you can identify yourself according to what you believe about God. If you are already familiar with the following terms, just check my spelling for corrections and move on.

Deist (deism): God is the creator of the world, but has stepped back and is no longer involved in our everyday lives. We're told that many of our founding fathers were deists.

Theist (theism): God is not only responsible for the existence of the universe, but takes an active interest in our affairs. God is a personal god who answers prayers and gives us guidance.

Atheist (atheism): One who holds the belief that there is no god. The universe came into being and continues to function without the aid of a supernatural spiritual power.

Agnostic (agnosticism): Agnostics don't know whether or not there is a god since they have not recognized any evidence of a god in creation or personally in their own life. Generously, agnostics are usually credited with the desire to find an experience of God. They look for evidence in the world or through reason and logic that is sufficient enough to convince them that God exists. Theologians tell us there are no, or very few atheists, that we are really mostly agnostics.

Perhaps there is no need to consider *pantheism* (God is in everything, rocks, trees, etc.) or *monotheism* (one god) versus *polytheism* (many gods). However there is another term that describes most of the theology of the Old Testament. Under the designation of the term, *henotheism,* there are many gods, but only one important God. Many scholars believe that this is the accurate description of much of the Old Testament.

We would have fewer atheists if we had a more mature concept of God in place of an image of an elderly white male with long hair and a long white beard who wears a robe and glasses. Even people of color may have been lulled into an image of a "white" god due to such cultural conditioning affected by the predominance of that image. One example is that 99% of cartoons depict God as white, standing on a cloud in heaven, being old and male.

Even though we have heard so often how we have created a God who looks human; yet we will persist in anthropomorphizing our concept of God. For those who may not be familiar with the term, anthropomorphizing is giving human qualities to something nonhuman or an object. To cite an interesting example, we need look no further than a parallel universe -- the Disney universe. No one seems to have noticed the dramatic difference between the two dogs, Goofy and Pluto. Pluto walks on all fours like a dog, doesn't wear clothes, and barks instead of talks. On the other hand, Goofy has been anthropomorphized. Goofy talks, stands upright on two legs, never walking on all fours, and is always dressed in human clothes. We have done that to God; we've given God human characteristics – created God in our own image!

Perhaps it is appropriate to assign human qualities to God. Or is it a matter of human beings inheriting God-like qualities: such as love, joy, and peace? And we certainly need to spend some time at some point in this section dealing with what it means for God to be personal!

It might be much better to start with a discussion of what God isn't. A cold impersonal definition of God would be the well known scientific suggestion that God is nothing more than the sum total of all the physical laws in the universe. Of course this is merely a simple semantical substitution. We have only changed the phrase "total of all physical laws" for the term, "God." Nothing else has changed. We know physical laws exist; no one challenges that. We just give that reality a new label.

To define "God" as merely the sum total of the physical laws of our universe, is essentially to say there is no god. On the other hand, a definition that describes God as a "person" or "three persons in one" seems totally incongruous with our universe.

We often attribute to God characteristics that are inappropriate. For example, I hear comments quite often that either make God seem inconsistent in God's relationship with us, or suggests God does cruel things that are totally contradictory with what Jesus reveals to us about God. God is said to be responsible for happenings in our lives. We hear the phrase, "God has a reason for this happening – we just don't understand why." Or we hear, "There is a reason for everything," most likely implying that God steps into our lives to control events. A little child dies, someone gets fired at work, a car accident kills three of the occupants but one survives, and we jump to the conclusion that God was responsible for the outcome because God "had a reason."

Let me illustrate how we blame God for very cruel things. Imagine a neighbor coming to your house and asking you for a special favor. He has a little eight year old girl and wants to teach her how dangerous it is to cross the street in traffic. His daughter has a puppy that she loves very much. To teach his daughter about this danger he wants you to get in your car. He will put the puppy across the street and will whistle and call the little dog. When the doggy runs out into the street to cross, he wants you to come down the street in your car and run over the dog, killing it. He said, this will be a dramatic lesson and will warn his daughter about the dangers of crossing the street. Such a dramatic lesson would teach her to be careful.

Simply reading about such an incident should have made you upset. You would consider this a terrible thing for a parent to do, even though we would agree the parent's motive was a good one: teaching the daughter to be safe. Good end – terrible and inappropriate means! It made me very uncomfortable just writing about it. And yet we often say God does these kinds of things to teach us lessons. We attribute to God behavior that we would consider criminal if we acted similarly. You're response may be that God is greater than us and "God's reasons are beyond our comprehension." I give a resounding "No such thing!" A loving God would never act that way! I call your attention to axiom number two that we agreed on at the start. The second axiom mandates that we rely on our own reasoning and value system. Which means we judge, to the best of our ability, the reasons that God may have for anything we attribute to God. Jesus described the love of God in very clear and reasonable language. This gives new and serious consideration to statements that we often use such as:

"We don't know why God took her"
"We are trying to make sense of why God would allow this to happen"
"We know God has a reason for everything"
"Tornadoes and floods are warnings from God"
"Someone up there is looking out for me"
"The man upstairs will take care of you"

The implication is while it seems un-godlike to us, it is because we cannot know God's reasons. It remains a mystery.

God is not responsible for those activities that we consider cruel. They are instead the evidence of human free will or natural occurrences of nature. God is not involved except possibly to provide comfort. Such comments, as immediately noted above, implies that God favors some persons and not others with assistance. This theology can easily excuse our careless behavior and cause us to not feel responsible for our choices and their consequences.

God is not inconsistent. Someone steps in front of a moving car; the car stops in time to spare the person's life. The next person steps in front of a moving car and is run over. Was God responsible in both cases? Did God choose to save one person and not the other? I believe the healthy answer is, "No."

Frequently we include God in athletic contests when we pray for victory before the event. It has become fashionable for athletes, after hitting a home run, making a touchdown, or making a winning basket to raise an arm and point a finger to the sky, implying that they believe God was a part of the success. You and I would agree that God does not have a favorite athletic team. If God had a favorite team, I suppose that they would never lose.

If there were no God, what would be the source of not only the universe, but such virtues as love, forgiveness, kindness, honesty and altruism? I remind you of our agreement to accept earlier stated axioms in Chapter 1. The second axiom was an agreement that we are permitted to discuss or argue and pass judgment from the perspective of our personal values systems. The above mentioned personal qualities such as love or honesty are not simply random words, but are in fact, very real and strong virtues or emotions that would not be automatically part of the physical universe. The existence of such personal qualities doesn't necessarily provide evidence of a personal God. Nevertheless, I am suggesting these virtues and other similar values give credence to a central spiritual law comparable to the physical law of the universe. Our universe runs on physical laws such as gravity, electromagnetism, etc. A compendium of spiritual laws govern our universe along with the physical laws. When we break the spiritual laws, we suffer the consequences. When we hate, it is destructive to our personalty; when we love, it enriches our lives. Such is the spiritual law. We need our spiritual values to fully understand our world and successfully achieve healthy relationships and community harmony.

If there is no God to maintain a set of spiritual principals, it forces us to struggle with the question of what motivates us to behave in a moral fashion. Why live a moral or good life in the absence of a supreme being and/or the existence of spiritual law? What keeps us from making pure selfishness as our guide besides the love of family and friends?

It does not provide an argument for the existence of God, but it feels comfortable and logical. Without a God to govern spiritual law, there will be nothing to discourage selfish and destructive behavior. The existence of our physical world and spiritual laws pleads for a designing creator.

Some Christian believers assign "intelligent design" to what we observe in nature and our total environment. (*Unfortunately, the term "intelligent design," is sometimes used synonymously with the terms, "creative design or creationism." It is often used by fundamental Christians to impose their religious beliefs and teachings into the public school systems. Religious teachings are inappropriate in publicly funded schools. This also includes all government institutions and functions. It is important to maintain a separation of church and state; religious teachings belong in our faith institutions, homes, religious schools, and our private lives. The only time religious studies are appropriate in public schools and universities is when all religions are studied as important historical components.*) We may have been too quick to totally ridicule the concept of an intelligent design for our universe. Criticism of intelligent design should not cause us to vacate our belief that our universe is more than just random. Evolution is the reasonable explanation for the way our world exists. This does not negate, but actually reflects the concept of an intelligent design.

We are so surrounded by marvelous wonders that persuade us to believe that it is all too magnificent to happen by chance. The human body alone overwhelms us with its complexity in nature. The heart that has kept us alive for seventy or eighty years; the wonderful work of our eyes; our special computer, the brain; and other body functions are incredible.

Science has revealed the impressive narrow limits of the connections of many physical quantities necessary to hold our universe together like the workings of a watch. Such relationships range from the extreme of large elements like galaxies down to the incredibly small elements that make up the atom. Chemical connections and physical relationships that hold our universe together operate in a very narrow window. The balance of many constants of nature must be infinitesimally fine-tuned for the various elements of the universe to hold together and work without chaos and for human life to exist. Such relationships involve delicate connections between temperature, gravitational attraction, and other components of various properties. Any deviation from the delicate balance which presently exists and our world would become uninhabitable. Our lives would be impossible if any one of several physical quantities had a slightly different value or increment.

Let me illustrate with an example from science. Consider the neutron mass that is only slightly greater than the proton mass. A reverse of the relationship between the two would cause the relationship to breakdown and make life impossible. The difference in strength between the proton and neutron is 0.2 percent. It constitutes a delicate balance or a fine line in the ratio of the relationship between these two particles.

Let me use an example that I understand. Sometimes I shave with an electric razor and sometimes a regular razor. In either case, the amount of blade that is exposed is very tiny. It has to be just the right amount to do the job. If the blade were not exposed enough, it would not remove the whiskers. On the other hand, if too much blade is exposed, it would remove more than whiskers and take off hunks of my skin.

Scientists tell us that the relationship between elements of this world, and particularly the forces of nature, is critical. For example, the strength of gravitational forces in connection with the motion of the moon orbiting the Earth must be just right to maintain a certain distance between the two bodies. Too much speed or too little and the gravitational pull becomes unbalanced. The distance between the two bodies will become less or greater and the moon will either crash down into the Earth or wander aimlessly through the universe until it might be captured by another planet, like Saturn or Neptune.

As we see, a careful examination of creation from atoms to galaxies reveals many critical relationships necessary to maintain an environment friendly to the evolution of our planet as we know it. Could this be evidence of the existence of a creative power responsible for our intelligent design. The debate between the intelligent design theory and the theory that the universe runs randomly with no purpose continues.

One of the reasons why a cadre of scientists are uncomfortable with the intelligent design theory is because it is an off-spring of the creation myth from Genesis, coupled with an anthropomorphic god. If we could come to agreement on a reasonable concept of some powerful force in the universe that gives purpose to existence, perhaps more scientists would be willing to endorse a creative force. It would be more than Spinoza's and Einstein's theories that the only god is the sum total of the physical laws of the universe.

Those with a discomfort with "intelligent design" generally find acceptable only a universe with no purpose. Our concept of the nature of our universe is that what we find here in terms of life and our environment is ubiquitous. Even if our universe is finite, there are still incredibly marvelous wonders to enjoy. It would seem a waste to postulate an environment with no people to appreciate the wonders. It seems unreasonable to remove people from the world and not have anyone to enjoy the wonders. Even though there are places where no life may exist to experience the environment, it is reasonable to conclude that intelligent life is scattered throughout the universe.

It is necessary to have the universe populated with various stages of life to make these wonders appreciated. Without this life, the rest of the universe is a waste. Consider the magnificent power of the ocean of waves crashing on a rocky shore. Or appreciate the beautiful black and orange of the Baltimore Oriole. Call up the image of a powerful and awesome polar bear. I hope you have had the opportunity to see the dramatic picture of the planet Neptune, a beautiful blue jewel as photographed by NASA, traveling through space. You may also have thrilled to the sounds of Tchaikovsky's Piano-Concerto #1. Something as humble as the iris, the snapdragon, or the gladiola is inspiring. We all enjoy the site of beautiful, puffy, white clouds in a startling blue sky. The following verse of a favorite hymn reflect these thoughts.

> God of the Sparrow God of the Whale
> God of the Swirling Stars
> How does the creature say Awe
> How does the creature say Praise
> (Words: By Jaroslav J. Vajda from what I call, "The Creation Hymn")

If you believe the wonders of creation are simply random accidents, consider science's description of how we came to be. As I shared in Chapter 2, astronomers tell us that every atom, molecule and cell in our bodies were a few billion years ago inside the body of the giant star. It exploded and

Cosmology and Theology

scattered these elements into space only to be coalesced over millions of years into our present bodies along with the formation of Earth.

Hopefully I have succeeded up to this point by sharing some astounding facts that may persuade us to reexamine an intelligent design involvement in creation. The beauty of our world is enough to be awesomely inspiring. Let's close with three numbers concerning creation that seem incredible.

First a very large number. Cosmologists tell us the distance from one side of the universe to the other side takes light approximately thirty billion years to traverse. That is at the unbelievable speed of 186,000 miles per second!

Now for a small number. Atoms are too small to be seen, but are made up of tiny particles called quarks. Physicists tell us a quark is so small that it is one hundred million times smaller than the atom.

Finally, a number more to our comprehension is the number of your ancestors. You have two parents, four grandparents, eight grand grandparents and the number continues to double with each generation. When you go back only thirty generations, you have over one billion great, great, great, etc. relatives to invite to your family reunion. The actual number, if you care to check my math, is 1,073,741,824.

We live in an astounding universe! Is it just an accident? Our existence and consciousness need to be present to give meaning to it all.

We mentioned earlier Einstein's definition of God as the sum total of the physical laws of the universe. An additional definition of Einstein's God as nothing more than the expression of human weaknesses was found in a 1954 personal letter. His letter was put up for auction with an opening bid of three million dollars. Actually your opinion in theology may be just as important and more accurate than Einstein's.

We cannot dismiss the personal experiences that people have claimed to have had of God. It matters not when anyone, including psychiatrists or psychologists like Freud, tries to analyze the many different and variety of personal experiences of God to the conclusion that such experiences are imaginary. To do justice and be fair, we simply acknowledge the possibility they constitute a true experience without any verification. People claim God spoke to them as clearly as we speak to one another. Others describe their personal experience with God as a special feeling that comes over them as John Wesley described one of his experiences. He said that his heart was strangely warmed indicating the presence of the Spirit of God. We must respect other people's experiences as possibly being real. As far as a personal experience of God, we can only deal with our own individual experience. Someone else's personal experience may not satisfy our own theology.

As we pursue the task of unraveling the nature of God, we look at what theologians and philosophers consider to be the most awesome and difficult question of all. This problem is what to do about a loving God and the existence of evil, pain and suffering in our world. Theologians pose the problem as either God is a loving God, but is incapable of helping us; or God has the power to shield us from pain and evil, but choses not to. God can do something, but doesn't; or God wants to help us, but can't. Is God good or powerful? It appears God can't be both. There have been many attempts to solved this dilemma. One book in the Bible that struggles with this issue is the book of Job. Some people believe that Job provides us a proper solution. Before we look at Job, let's consider other possible answers to this dilemma.

1. *It is necessary to have a world of contrast.* You cannot have an "up" without a "down." You cannot have a "here" without a "there." Contrast is necessary in order to have a three dimensional world. It would be impossible to even have a conversation without contrast. Language requires contrast. Imagine a piece of music with no contrast. A song with one note repeated over and over again would not be interesting. Even I can write that song. When it comes to the existence of pain and/or evil, a three dimensional world mandates that comfort, happiness, and all of the good things are balanced with discomfort, frustrations, unpleasantness. No "up" without a "down."
2. *Pain and evil make life challenging.* To say that they make life interesting is probably not a palatable idea. Life without tasks or problems to solve is a meaningless existence. We need a world that gives us important things to do and this means unpleasant

or unfortunate obstacles to conquer. We would not be much more than puppets without having to solve significant challenges.

3. *Pain and evil can be or should be character building.* Successful projects cause us to grow and mature into an adult-like state. There is a popular adage that claims that what does not kill you makes you stronger. It can be good for you to be roughed up or to face trials and tribulations. A walk through fire can purify us like refining gold.

4. *What we consider "bad" or "terrible" is only relative.* There are things that we consider horrible or atrocious because we see them from a limited perspective. When we get older and wiser, we see that what we thought was a bad experience (indeed pain sometimes seems unbearable), we come to realize these unpleasant experiences were exaggerated by our immature vantage. Let me illustrate from one of my personal childhood experiences. My mom and dad bought me a little stuffed monkey. I developed an affection for the monkey as little children will. There came a day when my father felt that I was old enough that I should give up my monkey (and perhaps learn to drive a car). It was not very funny back then. I looked out the window as my father took my monkey to the trash can and burned it. Just as many young children would find that a painful experience, my world came crashing down. Looking back on that unpleasant day, I can understand why I would feel so devastated. But as an adult, I can realize that my feelings were a child's view. In our mature stage what seems so bad is now not so devastating. This world can be very violent and painful. Someday from the perspective of heaven and eternal life, everything will take on an entirely new hue and color.

5. *Free will.* Consider the pain and suffering in our lives that is caused by human choices. Exercising free will makes life meaningful because we can choose between good and evil. Pain results from two sources; the natural world (tornado or flood) or those caused by humans (attacks, abuse, and other sins). In the later instance, God frees us to chose between righteousness and sin and our harmful choices can hurt others as well as ourselves. Free will is what makes us human. We must be free to choose between good and evil, or we are nothing more than automatons.

6. *We have a consistent world.* God is dependable or the world is dependable. We can count on risks and rewards, choices and consequences, decisions and responsibilities. According to the physical laws, if I jump off a ten story building, there are consequences. In like manner, any violation of the moral laws will bring harmful results. To choose sinful acts diminishes our spirits.
7. *Pain and evil make for a world with opportunity to exercise compassion and service.* It may not be comforting to you to believe that pain and suffering exist for us to have an opportunity to show compassion and love to others. But there are theologians and philosophers who site this as one of the possibilities. The world was created with conditions where we can help alleviate trouble and problems for other people. Such altruism is commendable and opens the door to love and kindness.
8. *Bonding with God.* Facing adversity and suffering in this world causes us to reach out for help. Many times people are driven to seek out God for answers and for help to deal with frustrations and a variety of other serious concerns. As a pastor I have had people come to me distraught and beside themselves. Faced with grief or other crises in their lives where they had difficulty coping, they would sometimes call at all hours of the day or night, confessing they had prayed to God for help and felt that God didn't answer. I often wanted to say to them that we don't wait for a crisis to occur and then reach out to God. We should be building a significant relationship ahead of time that would sustain us through the trouble. But that was the wrong time to give this advice.
9. *Returning to Job: the final answer is that we can't have all the answers.* There are mysteries about God and creation that we will never understand. The final solution at the end of Job is that Job must simply trust in God. Job wasn't around when God created everything. This is one of those moments where we must rely on faith and trust. (Reputable Bible scholars understand that the book of Job was not actual history, but a play or drama intending to address this difficult issue.)

To make more sense of this discussion, we need to package all nine points into a whole to provide some semblance of a solution for why God allows a world of pain and suffering.

The traditional view of heaven is a place of paradise where peace and joy reign. This contrasts with this life where there is pain and suffering. Do you remember the ending of Charles Dickens' *The Old Curiosity Shop?* Little Nell, the story's heroine, has just died after an unfortunate life full of pain and trouble. At the end of the seventy first chapter the text reads:

> "It is not," said the schoolmaster, as he bent down to kiss her on the cheek, and gave his tears free vent, "it is not on Earth that Heaven's justice ends. Think what it is compared with the World to which her young spirit has winged its early flight, and say, if one deliberate wish expressed in solemn terms above this bed could call her back to life, which of us would utter it!"

Now let's look at moral law.

Very few people would deny the physical laws of the universe. Scientists and theologians take as a given that we have a set of physical laws that keep our universe running. Gravity is the most familiar. This does not mean that all scientists attribute the existence of these physical laws to a creator God. Many scientists assume the physical law governing the movement of galaxies just occurs randomly. Our physical laws don't explain the source or the origin of all creation. Some scientists believe in a supreme being; some do not. You and I, and many others, believe there exists a set of moral or spiritual laws that govern our lives. The existence of moral laws may or may not contribute to evidence for a supreme being as the origin of such laws. It is difficult to imagine our physical laws just happening by accident. It may be even more difficult to propose a set of spiritual laws that govern our lives without there being a creator. It would seem the nature of spiritual laws almost mandates a spiritual source. Just as neglecting to obey physical law brings unfortunate consequences; ignoring or failing to obey spiritual laws would result in unwelcome ramifications. If we violated any one of our physical laws, such as ignoring the affects of gravity, we do so at own peril. For example, if you step out of an airplane without your parachute, you're asking for serious consequences, unless of course the plane is still on the ground. Even then you may sprain an ankle. Drinking a glass of poison violates a physical law with serious consequences. Likewise, breaking any spiritual laws will also result in serious repercussions.

The following list illustrates the functions of moral laws.

A. First, the function of moral law is to provide ethical guidelines. This law provides a behavioral structure that gives us directions for our decision making. We can explore the nature of such laws that determine appropriate and inappropriate behavior to help us avoid problems and make our lives safer.
B. The existence of moral law functions as a reward for appropriate behavior. Following the rules will help us build character and give our life structure.
C. Another function of moral law would be punishment or discipline for the sins we commit.
D. Spiritual law causes us to have compassion for others and provides the encouragement of such emotions as love, kindness, and thoughtfulness.
E. Finally spiritual law helps us bond and function in our relationships and the building of community.

Spiritual law expands our consciousness, allowing it to grow and develop. Let us consider some individual characteristics as examples of moral laws.

1. LOVE verses hate contrasts the most obvious repercussions for breaking a moral law. Love may be the strongest positive emotion as hate is perhaps the strongest negative emotion. (Although some theorists insist that hate is not the opposite of love.) The failure to obey the law of love provides the clearest and most dramatic example of the work of moral law. When we exercise love, our spirits or souls grow and develop, becoming more beautiful. When we express the emotion of hate, our spirit or soul decays.
2. Blessed are the PEACE makers. To be peaceful and gentle promotes harmony and cooperation. Failure to practice peace brings trouble and violence.
3. JUSTICE is an important positive characteristic that brings fairness and order. Ignoring justice leads to oppression.
4. WELFARE is our response to meeting the human rights of people in serious need.
5. HONESTY is essential in creating trust in personal and community relationships.

6. REVERENCE FOR LIFE is a very special virtue. I appreciate and applaud the person who cares enough about life that they take special precaution to protect living things. When I see someone abuse a dog or cat, or see a child take a stick and destroy a nest of ants for no good reason except cruel pleasure, I worry about that person's character. I believe it was Abraham Lincoln who said he pitied the person who could not feel the whip on another person's back. We will always identify "Reverence for Life" with Albert Schweitzer.
7. SACRIFICES people make by risking their lives to save others are most impressive. To offer to give up your life for someone else is the ultimate act of love. It emulates Jesus giving up his life on the cross for us.

These examples of moral laws at work suggests a source we might identify as God.

Three classic descriptions of the nature of God are omnipresent, omniscient, omnipotent. These concepts describe the ultimate supreme being. The definition of omnipotent is perfect power. Early in this chapter we dealt with the conundrum; either God is all powerful (omnipotent) and does not choose to protect us from evil, violence, suffering, or God wants to keep us safe but does not have the power or capability. This dilemma aside, we agree that the supreme being would be all powerful (omnipotent).

While the concept of omnipotence is clear, the other two descriptions, omnipresent and omniscient, seem too unbelievable in their implications. The concept of a supreme God being everywhere at once (omnipresent) would seem incredible. We are talking about a universe that is infinite or else finite. Bur it is immense in either case. How could God be present in every house, city, nation, or planet at one time. This is an example of what we mean by the inability of our minds to comprehend the greatness of God. We don't understand - we just believe. Early in this book we mentioned the concept of pantheism. God is present in everything. God, as omnipotent, is more than the total of all physical and moral laws. God is "being itself."

We come to the third classic description of God (omniscient) which states that God is all-knowing. A childlike view has God seated at a big desk with a large ledger open in front of "him", where the records of good or bad deeds are kept. God records everything (which would not be necessary for God would know everything). God would not need to write anything down. The idea that God knows everything is just as difficult to digest as the idea that God is omnipresent.

An interesting juxtaposition exists between God being omniscient and the concept of predestination. Predestination (mentioned in one of Paul's letters; book of Roman's chapter 8:29) means that everything was preplanned by God before it happened. God causes us to do the things we do; we have no choice in the matter. God planned everything from the beginning.

The interesting feature is if God were omniscient and knew everything, it would be tantamount to predestination which very few people accept. If God knows everything before it happens, it has to happen that way. Another conundrum! Is it possible to have God be omniscient, but not have predestination? Or does God being omniscient preclude or mandate predestination?

Let me illustrate. We know our friends and loved ones so well that we can predict their response to a phrase or event to a high degree of accuracy. For example, when I say to my wife, "wash your hands", I know her response will be an exasperated, "I was going to". This exchange has happened enough times that I am able to predict her response with ninety eight percent accuracy. From your own experiences, you could illustrate with many similar examples. The fact that I knew her response does not constitute predestination. If you and I can be that accurate in our predictions about such personal relationships, how much more accurate can an omnipresent, omnipotent and omniscient God be. But it does not constitute predestination. God is not a good guesser; God is a perfect guesser.

One of the problems we have difficulty avoiding is our image of God. We still cling to an anthropomorphic God. It is difficult to shake a concept of God as human. Comprehending such a mysterious supreme power is beyond our capability.

One closing thought relates to the grandeur of our universe. I want you to look once more at the cover of this book. You are looking back in space and time at a tiny portion of the universe that may be infinite and contain billions (if not a totally unlimited number) of galaxies. Each galaxy can contain billions and billions of stars, A significant percent of these stars with their accompanying planets could and may support various forms of life.

Whatever our theological position, all of us are left with a marvelous mystery. For example, no one has any idea what caused the universe to happen. It feels like the universe should not or cannot exist. On the other hand, it can't "not" exist because there would be absolutely nothing, including empty space. It does exist and will forever remain a mystery.

Are you familiar with the story of a little girl in Sunday school who was drawing a picture? The Sunday School teacher stopped by her side to admire her work and said, "Dear, that is a beautiful picture. What is it a picture of?" The little girl replied, "It's a picture of God". The teacher said, "But honey, we don't know what God looks like". Then the little girl said, "We will when I'm done".

4

IMMORTALITY

I never hear individuals talk about what I assume would be paramount on our minds as we approach our elder years. I would think we would wonder more and more about what is inevitable for all of us – death. When we are young, we do not contemplate death. Actually we can't imagine that we will ever be old. Such a condition is a lifetime away, in the dim unreal future. That may be one of the prime reasons why such messages as "smoking will shorten your life" are not very important to us when we are young. Such warnings mean little to people who can't conceive of ever being old – such as forty or fifty! However, I would be surprised if most people who reach seventy don't spend more time wondering just what awaits us after the moment of our demise.

Of course, speculating about our personal mortality may be uncomfortable or downright frightening. We may simply choose to refuse to think about it. I still believe that the majority of individuals do in fact worry or at least speculate on the subject as they age. The point of this last chapter is to use what we think we know about the nature of the universe and what we believe about God to extrapolate some idea concerning what lies beyond the end of life. Does the structure and truth about the universe give us any clues concerning what any possible existence after this life would be like? Does what we know about God help answer questions about life after death? Would any of this bring comfort to our anxieties concerning the subject?

As earlier when we discussed the nature of God, the question might arise, why not consult the Bible for our answers and thus remove all uncomfortable doubts? Then we would miss out on all the enjoyment of working through the process of developing our own personal theology. And, be honest, the Bible doesn't always give us clear and satisfying answers concerning "what lies beyond". For example, the teachings of Jesus describe a loving God which may not seem consistent with a concept of an eternal burning hell.

We should search for what seems consistent with the nature of the universe and a compassionate God, as we attempt to unravel what lies beyond death. THEN we can go to our Bibles and clear up our confusion! We go to our Bibles as an ultimate fact check. When we do turn to scripture, in Matthew 7:9-11, we find Jesus saying,

"Would any parent give a daughter or son a stone if they asked for bread? Or give them a snake when they ask for a fish to eat? If we who are imperfect know how to give good gifts to our children; think how much more a loving God will give even greater gifts!"

It would seem at times we attribute to God acts that are not even as loving as our own behavior. We cannot justify such thinking by saying God is so great we cannot understand the creator's actions. For example, the loving God that Jesus revealed is totally incompatible with an eternally burning hell!

There seem to be two possibilities for us after death. First, death could be the end of everything for us - the total destruction of our our souls, spirits or minds. Second, there will be some personal existence where we have an on-going self awareness or consciousness.

The first possibility is pretty straight forward. We just cease to exist, are totally obliterated forever, never again to regain consciousness. In the second possibility, there are various options: Heaven, hell, reincarnation and a variety of personal awarenesses.

The former possibility can be dealt with in short order. Our minds, spirits, souls are destroyed with our bodies so that we will never know how the future unfolds for those left behind alive on Earth.

The best thing that can be said about the first possibility of total destruction is that there will be no more pain or suffering.

Before moving on to the more encouraging alternative of some sort of consciousness or existence, we could dismiss the theory that the life of each person is totally destroyed by our suggesting that this would indicate that we live in a very cruel and godless universe. For example, there are people born into terrible circumstances, experiencing great misfortune, hunger and poverty. Other individuals may be handicapped for life, such as being totally paralyzed or dying at an early age. If that is the sum total of their personal existence – a few years of misery followed by an early death – how cruel is that? Only if compensated by something better after death, can one even suggest life is fair and not cruel. Contrast the long and comfortable life that many people enjoy with the life of the unfortunate, and the unfairness makes the universe seem absolutely vicious! But we believe that God is more compassionate than this. God has provided an existence beyond this life to compensate for any misery in this life.

Some philosophers will pontificate that when we judge the world to be cruel, we are imposing our human values. Yes, at times, they maybe wrong. We must trust that our values are practical and dependable. Though we could say our standards used in our evaluation of life and the universe may be biased, not relying on our own values leaves us disconnected with our world.

Since option one, the total cessation of all that we are at death, is not acceptable, the prospect of eternal obliteration would only be attractive to the deeply depressed who wish to "end it all" or to those who do not value the idea of a continuing life. If you wish to cling to such a dismal theory, it would seem you are discounting all beauty, love, joy, etc. of this life and are subjecting yourself to unhappy prospects.

Thus let's use Axion Two as one of the building blocks of our theology. The universe contains great joy, love, kindness, wisdom, relationships, adventure, and beauty. I believe that these experiences continue after death, revealing God's love for us and are clues that better things await us after death instead of total destruction. This is an assessment based on our human experiences and values.

Upon consideration, both option one—oblivion—and option two—eternal self-awareness—seem somewhat uncomfortable. No one wants to have it all end totally forever, unless that person is very depressed or totally unhappy with life and suspects that an eternal life would be no better. On the other hand, to exist eternally is an unsettling thought. Forever is a long time and I can't imagine a third possibility. "What do we do in the *forever time?*" is a unique question that we will consider shortly.

Let's explore the possibility that life does not end at death.

Arguably the most poignant statement against the existence of life after death is that it seems highly unlikely or very improbable. It is like a fairy tale, a magical or wild dream world. Some people may find it impossible to imagine. Remember the idea, taken from Lewis Carroll's <u>Alice in Wonderland</u>, *'I try to believe as many as six impossible things before breakfast.'* Simply reflecting on our world, we are astounded at the "staggering or awe-inspiring" things we must accept as fact or as unquestionable truths. Following is a list of wonders that we (or at least I) find very difficult to accept or understand. Yet they have the verification of science, the attestation of reputable witnesses or everyday experiences we use and take for granted.

For example, we all use the telephone (or did before we discarded it for texting). At one time or another you have used the phone to communicate with someone two or three thousand miles away. Unless you were unaware of what was actually happening you had to be highly impressed by the process involved. Our technology can take our voices, convert the sound to electronic coded signals, transmit it across a continent or around the world, convert it back to sound, in less than a second and so clearly that our voices can be recognized. As the sound of your voice approached your phone-friend, the signal was picked up by one of the millions of telephones (usually the right one) where your voice and exact words were reproduced clearly enough for the person you were visiting with to have no trouble understanding your message and recognizing your voice as she or he distinguished it from the many voices familiar to her or him. The communication happened at 186,000 miles a second and *it took less than a second to cross the continent*! If that experience doesn't stagger your imagination, neither will any of the following examples.

Imagine some passenger in a car picking up a cell phone and speaking into it, knowing it will be beamed through the glass of the enclosed car. Your voice will be carried perhaps several miles away to a tower and will be accomplished with a small battery. The signal is not a narrow beam, but a three-dimensional signal sent in every conceivable direction like a balloon being blown up and expanding. Impossible! The sound of the voice would be replicated so the person designed for the message would recognize that voice.

Recently I watched many beautiful photos on my television screen, pictures sent from a small camera which held hundreds of such pictures simultaneously. The images were each sent through a thin cable. That the camera held hundreds of these pictures and each picture was sent to the television screen through a small cord seemed impossible.

Similar to the above illustration is the electronic book reader available to us at a modest price today. These devices are the size of a single book but capable of holding hundreds of electronic books, and showing each one to you page-by-page. I say that is impossible from my perspective, and yet I see it happening.

Now consider the human brain and its capabilities. It has filed away billions of bits of information, the countless images of a lifetime to be recalled at will. You can conjure up a scene from a vacation taken half a lifetime ago, the dialog from a movie, a conversation with a friend ten years past, or events from a book you read. Our memory seems limitless.

As an example of the power of the human mind, a famous chess player, George Koltenowski, is known to have played 55 chess games at the same time. And he did this without ever looking at any of the chess boards!

Cosmology and Theology

Familiarize yourself about how the image of what we see with our eyes is transferred to our brain as an intricate picture. Our vision is light reflecting from the world around us, detected by the remarkable organs of our eyes, and thus perceived in all its details by our brains. Impossible, and yet you are doing it at this moment.

Some people find it impossible to believe that we sent a space ship to the moon and that humans walked on the moon's surface. They say it didn't really happen. It's too far fetched! And yet the astronauts even played a round of golf on the moon and returned safely to Earth -- to lie about their score.

As we observe the picture on the cover of this book once more, remember astronomers tell us the light from those tiny pin-wheel galaxies took billions of years to get here traveling at 186,000 miles a second. I repeat, billions of years!

Astronomy has provided so many illustrations of the awesome. That is the reason I used the picture on the cover. One example would be the existence of a pulsar star some 3,700 light years away from Earth which rotates on its axis 340 times a second. The diameter of the pulsar's orbit is 708,000 miles; and here comes the far out part, (no pun intended), astronomers have measured the light from this object 3700 light years away so precisely that they have determined *"the size of the pulsar's orbit changes by around a thousandth of an inch, less than the width of a human hair."* (Extreme Cosmos, Bryan Gaensler, Pages 65-66) (Italics mine)

Our natural reaction to the possibility of a spiritual existence beyond this life may also be one of amazement. I'm tempted to illustrate more profusely how incredible the life we are living is already. In contrast to the wonders of universe, it doesn't seem that an existence after death is any more startling than these miraculous things.

In light of all these amazing illustrations, it would almost seem more incredible if life did not resume after the transition of death. We are surrounded by these astounding realities in an incredible universe! All of the above is meant to impress you that the existent of life after death, impossible as it seems, is only consistent with the incredible things that surround us. So why is it so hard to believe that our being exists after the death of our physical body? It would be just another wonder in a world full of wonders.

That brings us to the reason for it all. It is no stretch of the imagination to claim that one purpose (or the main purpose of life) is to enjoy or appreciate life itself and all the aesthetic elements or experiences of our environment. Along with the "creative design" idea one can understand the magnificent enjoyments that thrill us: great music (Strauss waltzes, Tchaikovsky's sixth symphony, Chopin's music, Beethoven's sonatas), the beauty of nature (Yosemite, sunsets, irises, mountains, great waterfalls, the ocean), loving relationships of family and friends. It would seem a tragedy to lose such enjoyments. To experience such elements of life briefly only to have them disappear forever from our consciousness seems a terrible waste. The predication of an "intelligent design" creation would seem to indicate or include an extension of the best we experience in this life into a life beyond. As mentioned previously, some lives are tragically cut short or involve much pain and suffering. Thus life after death becomes a balancing of our experiences. It becomes a necessity to make life for all fair and just.

I often think it is sad to experience something special only to have it slip away into the past and be lost forever. We have our memories and, if we are prepared, perhaps we have captured the event on film or whatever technology exists for preserving happenings. However, memories and pictures are not as good as the original experience or event. And we have lost past opportunities to record that special event. We tape or record movies or programs off our television to view one or more times. The more interesting or exciting the program, the more times in our future we will watch it. There are many life experiences we would especially like to have for posterity. Imagine how unhappy everyone would be to hear of a great work of art (Michelangelo's sculpture of David, for example) being destroyed. Even more so the loss of a human life - a family member, loved one or friend - is devastating.

We can hope that in an afterlife, we could relive "those lost" times, special or beautiful experiences that we've had. On the negative side, perhaps hell is the reliving of the hurtful, embarrassing, or shameful moments that we wish had never happened or that we could completely forget.

Wouldn't it be wonderful to live through some outstanding historical events, hearing Abraham Lincoln deliver the Emancipation Proclamation, to be present as our founding leaders signed the Declaration of Independence, to have heard Dr. Martin Luther King, Jr. deliver his "I have a dream" speech or to hear Jesus teach.

For many of us there are special times in our lives that are just as important to us as the great historic events. They are very personal and are highlights in our life. They are so precious that we would love to find a way to reenact them. It is not enough to simply recall the special memories. Given an opportunity, many of us would choose to relive special events. For some of us it would be a wedding, a birth of a child, or our experiences that we would like to recapture beyond just remembering them. I am suggesting that we could have the power to relive the actual experience, not just remember it. Do you suppose heaven was created with these type of opportunities? If not, I would offer such a possibility as a good addition to the afterlife experience. We give meaning to the universe! It is our existence and appreciation of all of life that gives meaning to the universe and life itself. The universe would be a totally dark nothing without eyes to "see." Our world would be like Carlsbad Caverns when the guide turns out the lights for the tourists so that they might experience total darkness. The marvelous galaxies, stars, planets, etc. might as well not exist without life forms to experience them. There must be life throughout the universe. A universe empty of life is a terrible waste! We can extrapolate from that scenario to understand what the experience of eternal life must be like.

Continuing our discussion on the nature of eternal life, we must be conscious that living for an eternity is a long time! Without being factitious, we could ask the question, "What would we do with all that time on our hands?" Time extends from thousands of years to millions and billions of years and on into eternity which by definition has no end. Theologians sometimes tell us that there is no "time" connected or involved in eternal life; meaning there would be no sequential events. This takes us into the realm of the incomprehensible. We are unable to imagine the strangeness of "timelessness," and eternity remains impossibly long and frightening.

We are accustomed to having things to do to keep us busy. We have hobbies, work projects, and all sorts of activities to occupy our time.

Without these interests, we often become excessively bored and restless. So what will keep us occupied and busy in "heaven"? I can tolerate only so much harp playing. For one activity we could set about meeting and getting acquainted with all the other souls in "heaven" assuming most of us "make it." There are some seven billion of us presently on Earth. If everyone enters the eternal life experience, socializing eventually with all the others should keep us busy for some time. And this does not count the people who lived prior to our time or those who are yet to come. If we want to get really wild, there are an infinite number of creatures scattered throughout the galaxies pictured on the cover of this book. While getting acquainted with all these new friends, perhaps we will have the opportunity to visit the worlds from which they came.

Most people would like to think that an afterlife is a wonderful experience in contrast to this present life. An analogy, though weak, would be the *Wizard of Oz* movie. It starts off in black and white in Kansas, but changes to color in Oz.

We know our world through sight, sound, smell, taste, and touch. We know other people through speech and ideas. Our senses require a physical body. How do we communicate without bodies? Will we exist in a spiritual dimension where there is a significant sensing with abilities comparable to our five senses? It seems necessary that there be an ability to sense our environment and communicate with one another in order that we would find it a significant and satisfying experience. (No doubt English will be the official language in an afterlife!)

If we do not use our five physical senses to experience our environment in a purely spiritual state in an afterlife existence, it poses a conundrum. Mark Twain addresses this in *Captain Stormfield's Visit to Heaven*. Two of Twains favorite subjects were cosmology and theology just as mine are in this book. He spent an inordinate amount of time meditating on God, heaven and sin, as well as astronomy. In his writings, he has his character after death flying through space on his way to heaven. When the captain reaches the sun, he passes through it without any injury because he is a spirit and impervious to physical harm. While Twain is cognizant of the inability of Stormfield's spirit to sense physical fire or heat; he forgets that Stormfield would not "see" the bright fire of the sun he was passing through. (Incidentally, in the popular television situation comedy, *The Big Bang Theory*, we learn that Superman cleans his costume by flying "unharmed" through the sun.)

We might speculate that relationships, and thus communication, would be a necessary part of an afterlife experience. As we consider our ability to communicate, let us consider the mind that is connected to the soul or spirit. Our understanding of the brain – not just the physical organ, but intellect – confuses mind, spirit and soul. One definition of *mind* is the aggregate of all conscious and unconscious processes – our thinking, awareness and memory. *Soul* can be described as that part of the *person* that survives death – the immortal force or essence. We often conflate "soul" and "spirit." Some people believe that soul, spirit and mind are linked together in a mysterious way that can be separated from the organ of the brain. Many scientist believe there is nothing apart from the brain – that soul and spirit don't exist. They say that the mind and thinking are simply "chemical" processes of the physical organ.

To challenge this idea that our brain activity is only chemical or merely physical processes, let us consider the capabilities of our brain and what is really happening when we think. We call up anything we choose to entertain us or think about. Scientists maintain that what we think about is triggered by some outside stimulus. We see an object or someone says something that initiates a chemical sequence of memory, causing us to follow a chain of thought. Actually you and I are able to call up anything we choose without the presence of any outside stimulus. We can direct our thoughts any way we determine. We can decide to recall a movie we saw or some place or event from our childhood without anything to trigger the thought. If I suggest that you use your creative volition right now, you can pursue any flight of fantasy. Scientists say your flight of fancy was triggered by my suggestion. However, I only suggested what you should try. Your choice of subject is a product of your free will.

An article in <u>Scientific American</u>, "The Language of the Brain," (October 2012 special report issue, pages 54-59) claims that "Neuroscientists do not fully understand how the brain manages to extract meaningful information from all the signaling that goes on within it." The brain is "capable of perceiving, thinking and acting with a finesse that cannot be matched by any computer."

Some philosophers make a case that we have no free will, but that all choices are determined by logical connections of sequences in the brain - that every act is influenced by previous acts. I don't believe that works here. Self experiment proves that we have the potential to exercise free will. As we just iterated, you are able to call up any past event or experience you choose, focusing on any details you want.

How does cloning enter the picture? I believe that there would be no transfer of memory from one creature to its clone. Might this be possible evidence of separation between the physical brain and the spiritual existence?

What scientists are suggesting is that all the processes are simply chemical reactions and nothing more. This seems too impersonal and mechanical to be the sole mechanism of our minds, our consciousness or our self-awareness. I prefer the view that the soul, mind or spirit is a separate existence apart from the body. There is a mysterious connection with the body in this physical life that terminates with death when the spirit is released.

In my earlier discussion of the universe, I denigrated the idea that there are extra dimensions apart from the three dimensions of our existence that we experience. I later contradicted myself by suggesting that God exists in a separate dimension. Now I further contradict myself by saying that the spiritual existence after death constitutes yet another new dimension. Perhaps God and an afterlife exist in the same dimension. Let me give you an illustration that might be helpful. The continuum of electromagnetic waves extend through the ultraviolet (short wave lengths) dimension that is invisible to our eyes into the visible portion of the continuum and back into the invisible infrared (longer wave lengths) portion. The visible portion of the light wave could represent the physical life while the invisible portion at either extreme of the continuum represents the spiritual dimension.

Only peripherally related and parenthetical to our subject is a new possibility. The capability exists where a funeral director extracts the carbon (the prehistoric vegetation that over millions of years has become coal) from our bodies and changes it into a synthetic diamond for our loved ones to wear.

If after death we continue to exist on a spiritual plane where our five senses are not applicable, are there alternate senses that we might use to communicate or navigate through our new spiritual dimension? How do we communicate with each other and how will we recognize each other? What seems more bizarre, how do we even find one another? When I die and become a spirit and I want to have a visit with Grandpa, John Henry Terry, how and where do I find him? What if he is off exploring Saturn or in another galaxy far, far away? I suppose the easiest solution will be that I can simply "text him." Perhaps he would not be wondering around the universe, but will be waiting "on the other side" for my arrival. Now we are using concepts, "time" and "space" which may be irrelevant.

Now let us consider reincarnation, the idea that when a person dies, their spirit is reborn as a new person at a different time. When the new life begins, there doesn't seem to be any connection with that spirit and the family and friends from the previous life. This makes some people, and me included, terribly uncomfortable with reincarnation. Perhaps I am too sentimental, but I feel that the loving relationships we build in this life are special and beautiful. They will continue on into our existence after death, so that post-death will be meaningful. "Heaven" has to include our loved ones and friends for it to be "Heaven." If there were any connections with people from one life to the next under the auspices of reincarnation, we seem to be oblivious to our association with those from our past lives. I'm sure that you loved them too much for that to be the case.

The main purpose of this chapter is to examine some evidence for an afterlife and what it will be like. If there is a human spirit that transcends death, it raises some interesting, but troubling questions. 1) What happens to the soul or spirits of people that are never born? 2) When do prehistoric people reach the stage where they are human? 3) What is the fate of animals?

The first question relates to a possible potential for there to be more people that never will be born. A family may have a different number of children, but have the potential to have more. Each of their children is special and has a unique personality and life all their own. The question is, "What happens to more children that they could have had, but didn't?" There is a loss from those children who weren't conceived but must have a potential existence and personality waiting to be born. If they are a potential existence that never happens, what becomes of their "life?" Will they be born to another family? Do they remain in some spiritual state? Do they go directly to an afterlife? These are extremely difficult questions that I have no answers to suggest. They are similar to the complex questions raised during the philosophical discussion of the nature of space and time that bewilders us. I'm also not attempting here a discussion of potential lives that are never brought into existence in connection with birth control or abortion.

A second troubling question relates to evolution. Assuming evolution accurately explains how life developed on Earth, at what point did our ancestral species evolve from animals without a soul to people with souls? At what point did we qualify as a candidate for "heaven?" Is the demarcation before the Neanderthals or is it later in biological history? Geneticists, evolutionary biologists and anthropologists tell us the human Genome has changed less than 0.02 percent in 40,000 years. Was there a moment in time when a first creature stepped through "that door," developed a soul and became human?

The third question concerns animals. Do they qualify for an existence after death? It will seem appropriate that they have a capacity for existence after death for the parallel reason that people who have had a miserable life are recompensed with an after death experience to make up for a difficult life. Do domesticated animals have a "seed" of advanced existence because of their relationship with humans? Can we truthfully say to a little child, "Don't cry, Honey. Our pet is gone and now waiting for us in Heaven."

Do you remember reading the account of a cat rescuing her kittens from a fire? She returned to the burning building again and again, suffering severe burns while rescuing her kittens one at a time. The newspaper article captured the imagination and sympathy of the public. After the amazing rescue, it was sad to read that the kittens were taken from her and adopted out to various homes. Did the mother cat suffer the loss of her kittens? I believe that pets have feelings, even if not to the extent of our feelings. I would like to think that abused animals have some reward to make up for their suffering.

Before moving on from animals and afterlife, it is interesting to note the attitude that many Native Americans have traditionally had toward animals. Native Americans asked forgiveness from the animals that they killed for food -- an attitude towards creation that all life was sacred. This is similar to Albert Schweitzer's theology of a "reverence for life." He was a great missionary to Africa whose theology posed a dilemma for him when he rescued an injured bird that needed to eat live mice to survive. He had to choose between the life of the bird or mouse.

Returning to our main theme, we ask the question, "If there is some comfort in "heaven" to compensate us for any suffering in this life, what about the punishment we deserve for our bad behavior? Almost everyone subscribes to some idea of fair play. When we are naughty or commit sinful acts, there should be a consequence. You may suggest that the Bible has allowed for our deserved punishment with the existence of a hell. We discussed this in the previous section on the nature of God. To believe in eternal torment for the wicked doesn't match what Jesus taught us concerning a loving and forgiving God. Is your sense of justice offended by the idea that there is no hell? Let me suggest an alternative to replace the extreme of eternal torment.

Each one of us hopefully recognizes the evil we do or the sins we commit and have feelings of remorse. If you observe me walking down the street and noticed that I just paused to clinch my fists and flinch, it may be because I was just reminded of something bad I did or some embarrassing incident. After I die, and you are surprised to see me in heaven, let me assure you that I'm being punished for my sins. I will be experiencing remorse because my loved ones and friends finally recognize all of my weaknesses which were hidden from them in this life. Do you have any confessions that you want to make and get an early start on your feelings of remorse?

If one of the purposes for us in this life is to grow and mature, shouldn't this process continue in the afterlife then as an extension of this life? In our lives we are busy collecting knowledge, establishing relationships, pursuing objectives and goals; why not continue these adventures in "heaven?" We could learn additional languages or whatever our form of communication would be; explore more wonders of the world as well as other worlds in the universe; write a book or whatever form expression takes in the new realms or become more knowledgeable concerning history. If collecting is an option, we could complete our stamp collection or other hobbies. We hope our mental skills will be sharper and to some extent, unlimited in their capacities. That would mean that we can become better chess players, but so will our opponents.

For some people a bizarre, but related, subject is ghosts. Are ghosts evidence of the ongoing life after death or simply a figment of our imagination? We would like to believe in ghosts if ghosts indicate immortality.

In the popular story of Peter Pan, our hero finds himself standing on a rock in the Mermaid's Lagoon with rising water that threatens to drown him. He is afraid of dying, but suddenly is encouraged by a message of some inner assurance, causing him to exclaim, "To die will be an awfully big adventure."

www.ingramcontent.com/pod-product-compliance
Lightning Source LLC
Chambersburg PA
CBHW021005180526
45163CB00005B/1904